The Highest Stakes

The Economic Foundations of the Next Security System

WAYNE SANDHOLTZ • MICHAEL BORRUS
JOHN ZYSMAN • KEN CONCA • JAY STOWSKY
STEVEN VOGEL • STEVE WEBER

A Berkeley Roundtable
on the International Economy (BRIE) project
on Economy and Security

OXFORD UNIVERSITY PRESS
New York   Oxford

Oxford University Press

Oxford   New York   Toronto
Delhi   Bombay   Calcutta   Madras   Karachi
Kuala Lumpur   Singapore   Hong Kong   Tokyo
Nairobi   Dar es Salaam   Cape Town
Melbourne   Auckland   Madrid

and associated companies in
Berlin   Ibadan

Library of Congress Cataloging–in–Publication Data
The highest stakes : the economic foundations of the next security system /
Wayne Sandholtz . . . [et al.].
p.   cm.
"A Berkeley Roundtable on the International Economy (BRIE) project
on economy and security"—
Includes bibliographical references and index.
ISBN 0–19–507035–6
ISBN 0–19–508667–8 (pbk.)
1. United States—Military policy—Economic aspects.
2. United States—National security—Economic aspects.
3. United States—Economic policy—1981–.
I. Sandholtz, Wayne.   II. Berkeley Roundtable
on the International Economy.
HC110.D4H54   1992   338.973—dc20   91–43270

2 4 6 8 10 9 7 5 3 1
Printed in the United States of America

A Berkeley Roundtable
on the International Economy (BRIE) project
on Economy and Security

The authors are grateful for the support of the
MacArthur Foundation and the Sloan Foundation.

Acknowledgments

The Highest Stakes assembles into a single book the research of seven scholars extending over many years. The undertaking involved both a joint research project and an extended intellectual debate and argument within the group. It began as a research project some seven years ago, bringing together graduate students and faculty at Berkeley. A series of research monographs was the product of that initial phase. In the second phase, the monographs became three dissertations and a series of publications. Several additional research efforts were also launched to complement the original work. In the final phase, other researchers were brought into the debate. In the end, these chapters have been crafted from our collective work into a single argument, not a conventional, edited volume.

Throughout its long gestation, this book benefited from the advice of many wise and wonderful people. Thanks go to our BRIE colleagues, François Bar, Stephen Cohen, and Laura Tyson, for providing extraordinary levels of collegiality and intellectual support. We are also collectively indebted to the following diverse group of friends for comments, criticism, and encouragement (and to those who know their contribution but may have been inadvertently left off the list): Ken Adachi, Suzanne Berger, Jamais Cascio, Manuel Castells, Marcello de Cecco, Jim Fallows, Peter Gourevitch, Sid Hecker, Stanley Hoffman, John Holdren, Cordell Hull, Takashi Inoguchi, Chalmers Johnson, Robert Keohane, Kunio Kinjo, Yasuo Komoda, Jim McGroddy, Ruben Mettler, Hideo Miki, Joseph Nye, Victoria Rehn, Gene Rochlin, Jean-Jacques Salamon, Masakatsu Shinkai, Etel Solingen, Charles Sporck, Audi Steele, David Stuligross, Nozomu Takasaki, Greg Treverton, Sachio Uehara, Felicia Wong, Nicholas Ziegler, and the Scripps College Faculty (for research funds for Chapter 3). Special thanks are due all the other people who gave their valuable time to be interviewed or otherwise aided the manuscript's preparation.

A special debt is owed to BRIE's Joseph Willihnganz who singlehandedly turned an unwieldy manuscript into a tightly-edited whole. Indeed, the book's production would have been impossible without the professionalism and infinite patience of BRIE's staff.

The project and the book were made possible by principal funding from the MacArthur Foundation and the Alfred P. Sloan Foundation. The conversion from separate essays into a book was facilitated by the cooperation of Oxford University Press, especially by David Roll, Mary Garrison, and Herb Addison.

Finally, we would each like to close with a personal dedication:

In loving memory of my father, and for my mother.   Michael Borrus

To Tina for her patience.   Ken Conca

To Judy and Sarah.   Wayne Sandholtz

For my parents—my first teachers and still my best.   Jay Stowsky

To my wife, Susan.   Steve Vogel

To Felicia Wong.   Steve Weber

For Vikki and Lara.   John Zysman

Contents

I

THE STAKES

Prologue

MICHAEL BORRUS, WAYNE SANDHOLTZ,
STEVE WEBER, AND JOHN ZYSMAN

The economic foundations of the postwar security system have eroded. Industrial and technological initiative in Europe and Japan is creating the basis of a wholly new system that minimizes U.S. influence. Well before the opening of Eastern Europe and the collapse of the Soviet Union, a redistribution of global economic capabilities was already defining new arrangements of power.

When politics catches up with the new arrangements of economic capabilities, the world will look different, perhaps very different, from how it looks in 1992. This book explores how the industrial might of Japan and the increasing economic unification of Europe might affect the politics between nations. The redistribution of economic capabilities will reshape the international political system of relations, but will not uniquely determine the shape of the international political system. In 1945, American economic preeminence and industrial leadership certainly ensured that Washington would be the leader of the West. They did not, however, determine the character of the system that Washington shaped. There were many choices then, and there are even more choices today.

Our aim is to demonstrate how the economic and technological changes in the major industrial powers will shape these emerging choices in the international system. For that reason, this is not an argument about the relative decline of the United States, nor necessarily an argument about marginal changes to sustain the American system. Instead, it is a story about changes in power and interests, changes that may be important enough to presage a new security era—an era in which military threats to territory and society recede, only to be replaced by new, more sophisticated ones; an era in which "security threat" no longer refers just to tanks and missiles but also to the control of markets, investment, and technology; an era that recycles old security vocabulary to fit new issues: market share, protectionism, relative gains from trade. An era, simply, that would reconceive the very character of security, redefine the international power game, and resituate its players.

The collapse of the Soviet threat throws open the international system. Events of the past two years—Soviet withdrawal from Eastern Europe, evisceration of the Communist Party, and the power shift from Moscow to the republics—have removed the last motive for seeking shelter under the American umbrella. Conflict within or between the republics would have international repercussions, but it will not pose a threat of aggression in Europe or Asia. With that gone, the reconstruction of the international security system will depend more directly than ever on the relative economic power and influence of the United States, Europe, and Japan.

The new distribution of economic power ensures that the system will change dramatically. By the 1980s, Japan and Western Europe had developed independent technological, industrial, and financial strengths, and the U.S. economy was hegemonic only in memory. Allied dependence on U.S. leadership was already more a matter of U.S. sunk investment in military capability and allied unwillingness to redirect economic resources—in short, more political choice than economic necessity.

The first part of the book sets the stakes—the emerging character of the international system, and the place of the United States in it. Chapter 1, "Industrial Competition and American National Security," argues that the exercise of American power is undermined by the loss of technological and industrial leadership. The problem is not that U.S. foreign policy commitments have exhausted the economy's resources, nor is it that others have simply caught up. Rather, in a variety of ways, Japan and Europe have innovated, creating distinct regional foundations for new roles in the international system.

The second part of the book considers the players. Chapter 2, "The Power behind 'Spin-Ons': The Military Implications of Japan's Commercial Technology," examines the security implications of Japan's industrial prowess. Chapter 3, "Europe's Emergence as a Global Protagonist," explores how Europe's latest efforts at economic and monetary integration are redefining Europe's political identity and its role in the world. Chapter 4, "From Spin-Off to Spin-On: Redefining the Military's Role in American Technology Development," returns to the American case. It examines the Pentagon's reaction to new industrial developments abroad. It argues that promoting defense technology will not restore our technological capabilities. Rather, traditional defense spending risks undermining both our commercial and our military position. Chapter 5, "Third World Military Industrialization and the Evolving Security System," contends that military industrialization in the Third World will not alter the international economic and security systems. The major players in this game will remain the advanced industrial countries.

The third part of the book speculates on the character of the changing game. Chapter 6 explores "The Risk That Mercantilism Will Define the Next Security System." It proposes that the current openness can be sorted into at least three possible futures. The system of relations that will emerge among the great powers could, and we hope will, become *managed multilateralism*, an adjusted version of what we are in now. In this case the security problem dwindles or vanishes. But that is only one possibility and should not be taken for granted. The emerging distribution of economic capabilities also suggests two possible forms of regionalism as alternatives. *Benign regionalism*, a sort of defensive protectionism, projects an international economy composed of three substantially self-sufficient regional blocs. Cross-regional ties would be less important than intraregional ties; the regions would interact, but there would be limited cause for disagreement. By contrast, the third alternative is *regional rivalry*, a twenty-first-century form of mercantilism, where the drive for autarky would be driven not by welfare concerns but by aggressive, beggar-thy-neighbor economic strategies focused on accumulating state power at the expense of others. Security in that world would be a very different game from what we have become accustomed to over the past forty-five years.

1

Industrial Competitiveness and American National Security

MICHAEL BORRUS AND JOHN ZYSMAN

The debate on U.S. competitiveness must become a debate about national security. Relative decline in economic position and failing technological leadership could soon undermine the exercise of American power. Many argue that extensive foreign policy commitments have exhausted the economy's resources.[1] Others point to an increasingly sophisticated and interconnected world economy, accepting that U.S. security must now rest, in part, on foreign technological and industrial capacities that lie outside of direct U.S. control.[2] But these views fall short of considering the most fundamental transformation: American industrial and technological decline has eroded the foundation of the postwar security system. At the same time, industrial and technological initiative abroad is creating the basis of a wholly new system that could markedly reduce U.S. influence.

These concerns may seem laughably distant coming on the heels of America's remarkable military victory in the Persian Gulf and the collapse of the Soviet threat. To be sure, the Gulf conflict demonstrated the vast difference between a great power and the modest capabilities of a regional Third World power. The United States was able to mobilize the resources necessary to alter events half a world away; it proved it could still project its will into a distant regional conflict. The combination of America's sophisticated electronic weaponry, command and control systems, and military strategists proved utterly dominant against a technologically and tactically overmatched adversary.

But U.S. military success in the Persian Gulf rests on past industrial strength; it is not a reliable indicator of future capacities. Even American weapons mastery rests on electronic components and subsystems largely designed in an era when U.S. industry dominated the civilian computer and semiconductor industries. That era is fading rapidly. Continued mastery is by no means assured because the economic base on which the U.S. international position rests is at risk.

For four decades, the international security system presumed a fundamental

Soviet enemy, a U.S.-controlled military umbrella over allies in Western Europe
and Asia, and an international system of trade and finance (excluding the
Soviet bloc) dominated institutionally and in the market by U.S. economic
strength. Each of these pillars of the postwar security system has changed profoundly: The Soviet Union has collapsed; the American provision of military
security for allies in Asia and Western Europe, once an economic and technological necessity, now endures only as a mutual political choice; and, finally,
U.S. economic hegemony has been challenged by dramatic shifts in industrial
and technological position. This third pillar, the changing economic foundation
of the security system, is the central concern of our analysis.

American economic capacities have always been more limited than commonly perceived (creating, as Samuel Huntington has remarked, a "Lippman
Gap" between extensive commitments and more limited resources).[3] Not that
those capacities are insignificant: The U.S. economy remains the world's
largest, and its technological and scientific resources are still deeper and
broader than any near challenger in Europe or Asia. But the relative U.S. position has changed substantially. Faster growth abroad was to be expected as
countries such as Japan, Germany, and France rebuilt their domestic
economies, borrowing the best industrial practices, licensing technologies from
the United States at modest cost, and successfully shifting resources out of agriculture into industry. But catch-up abroad did not simply restore a more traditional balance of economic power—the differential in growth rates continued,
driven, in Asia especially, by high rates of productive investment. Today, the
absolute level of industrial investment in the United States has fallen below that
of Japan, though Japan boasts only half the population and GNP of the United
States.[4]

Profound shifts in the United States's position in technology, trade, and
finance will not be easily reversed. American technological leadership is now
severely threatened in a range of important areas, including electronics. Increasing numbers of U.S. industries now retain their market position only through
heavy doses of trade protection. And the United States has moved quite suddenly from its position as the world's largest creditor to become the world's
largest debtor.[5]

The United States is at once more vulnerable to, and constrained by, decisions made abroad, and it is less able to exert its influence on behalf of foreign
policy objectives. There is little disagreement about this despite widely varying
interpretations of the American decline.[6] The issue, then, isn't whether the relative U.S. position has changed, but what significance to accord its decline.[7] In
our view, the change in American industrial position is hugely significant: It
augurs a transformation of the international security system by upheaval at its
economic foundation.

New Economic Dimensions of Security

Historically, the principal concern of security systems has been the control of
territory and resources.[8] For a nation, the question has always been how to

preserve the community within national territory from outside intervention and control, while pursuing the community's shared goals in the external world.[9] Control and use of resources has traditionally required armed force. Now, however, private actors can control significant resources through markets, while governments influence their operation for national purposes. The security issues do not disappear, but they become submerged and hidden by market relations. As the threat of military conflict among the advanced countries dwindles and the influence of international markets grows, the question of how to achieve security must then become twofold: (1) How can a nation maintain security in the traditional sense of guarding its borders and resources? and (2) How can a nation preserve its economic integrity and social community—or differently, how can a nation's aims be preserved in the constraining web of international economic relations? Suddenly, market structure and market function become a matter of intense concern.

We may consider a nation's economic security in terms of its ability to generate and apply economic resources to the direct exercise of power, or to shape indirectly the international system and its norms. When allied nation-states are knit together into a shared security system, the power within the alliance resides with the nation that has the ability to get the others to act on its behalf or the wherewithal to put to its own use resources belonging to the other states. That can be accomplished directly through overt threat and punishment (or promise and reward), or indirectly when the structure of the system produces outcomes that serve the lead state's interests, or when the lead state's preferences become the alliance's accepted norms.[10] In the postwar security system, the economic dimension was critical to the direct exercise of U.S. power. Industrial and technological resources supported U.S. military strength and underwrote the use of commercial and technical assistance to secure allied agreement with U.S. goals. And the economic dimension was just as critical indirectly in its impact on the system's structure and norms.

The economic structure of the postwar security system rested on multilateral free trade established by the General Agreement on Tariffs and Trade (GATT) and international financial stability embodied in the Bretton Woods and successor agreements.[11] After the war, the United States generally kept its market open while tolerating Europe's and Asia's departures from free trade for the sake of their development. The United States, in fact, encouraged their development through extraordinary transfers of finance and technology. In turn, the United States benefited by selling technically advanced goods and services that others could not provide and then purchasing foreign assets with the surplus profits and overvalued currency.[12] In this way, by both rewarding allies and extracting tribute from them, the United States channeled compliance for its policies in the security realm.[13] Directly and indirectly, then, the U.S.-led security system has rested on industrial strength and technological leadership.

Now, from a position of dominance, America has begun to slide—risking dependence in industry, finance, and critical segments of technology. American industry's inability to adjust competitively to changes in global markets threatens to undermine the U.S. commitment to openness and the ability to achieve U.S. goals. America's ability to exact allied compliance, either directly or through structuring of the system and its norms, has diminished. Conversely, the allies need no longer covet U.S. technology or financial resources. In an increasingly large range of commercial technologies with significant defense application, Japanese components and subsystems and European manufacturing equipment and materials are broadly equivalent to or even superior to American ones. The restructuring of Europe and the continuing rapid growth of an Asia dominated by Japanese economic resources reinforce the probability of autonomous regional actors and create the possibility of alternative regional security arrangements.

The Argument

This chapter builds the case that a transformation of the international system is under way and requires a rethinking of the basic concepts of security. The first section proposes that America's decline relative to its allies is fundamental, the result neither of catch-up abroad nor of imperial overreaching. Rather, it is the basic loss of industrial position that is undermining the economic and technological basis of the old security system. The second section examines the emergence and consequence of powerful new industrial capabilities in Japan and Europe.

Section III considers alternative technological bases for security by examining the changing ties between military and commercial technology. It suggests that military technology cannot rescue the commercial U.S. position—but commercial weakness can undermine military strength. Section IV explores whether or not the emerging multipolar economic system will produce a multipolar security system. The American, European, and Asian regions each have the political capacity and techno-industrial foundations for independent action. There are, moreover, disturbing indications that increasing regional autonomy will introduce new and severe constraints on U.S. policy—the kind of constraints that the United States has become accustomed to imposing, not accepting.

The fifth and concluding section argues that the new technological and economic foundations that are emerging could support any of a number of security structures. The distribution of economic capabilities does not, however, dictate the precise form of an alliance system. The alliance system will be a product of varied conceptions of strategic order, the national and domestic group interests at stake, and the particular crises that force old arrangements to be reconfirmed or new arrangements to be created.

Our analysis stands in marked contrast to the intellectual and policy debate now being formulated in response to the fragmenting of the former Soviet Union and the instability of the Persian Gulf. The emerging debate presumes that as American hegemony wanes (and even this is questioned to some extent), the security system will evolve toward mutual interdependence with continued

U.S. leadership.[14] The implication is that the existing economic structure of the security system—the distribution of capabilities—will also support mutual interdependence under U.S. leadership. In this view, a multilateral security system can be successfully managed by the United States. "Managed multilateralism" can retrench America's security position to a defensible point consonant with its relative economic power. A repaired and strengthened system of open trade and stable finance can, in that view, continue to structure outcomes favorable to American interests. This would certainly be a preferred outcome, at least for America, of the current upheaval.

However, to presume interdependence as the natural successor to American hegemony misunderstands the processes driving the evolution of the international economy and their significance for the U.S. position. To be sure, the presumption of interdependence is supported by the increasingly global character of major firms. But it fails to acknowledge the stubborn reality of regional markets and local industrial communities, as well as the enduring role of national and regional policies. The temptation to choose between the global and national phenomena must be avoided; we must not deny one evolution by pointing to the other.[15] Rather, the task is to sort out the interconnections of economic activity at the global, regional, and national levels—and then to understand the implications for security. In fashioning a new economic basis for security, the United States faces more than the simple choice between defending free trade or succumbing to nationalism. We must not view our options as an either/or proposition: either we strengthen general rules to sustain an independent multilateral system, or we act bilaterally to advance our position in a world of competitive regions. We must pursue a measure of both these strategies.

I. The Emergence of Vulnerability: America's Deteriorating Position in the Global Economy[16]

How deeply eroded is the American capacity to exact compliance either directly or through its position in the trade and financial system? There are two interpretations. One view is that the U.S. decline is mostly the result of the industrial catch-up of Europe and Asia which ended in the mid-1970s. The second view, presented here, is that the decline is more fundamental, and has been disguised by the process of catch-up, stagflation, and European economic troubles in the late 1970s and early 1980s.

To assess these competing positions gauging the depth of American decline, we must consider more than aggregate growth or trade figures. We must consider the evolution of industrial technology and the dynamics of international competition—an evolution that suggests that international markets for technology, manufacture, and finance no longer unquestioningly support U.S. industrial leadership. We must understand that the ability of the American economy to adjust to shifts in international markets and to the emergence of new competitors has substantially diminished. We must recognize that America's external debt is now a potential constraint not only on foreign policy, but on fiscal policy—a point that was brought home by recent contentions that continued American pressure for financial market deregulation would provoke a cut-off of Japanese credit. We must realize that all these developments represent new and serious constraints on U.S. power.

The Pattern of Decline

We see the overall picture like this: An emerging competitive weakness in manufacturing, increasingly visible in the 1970s, was accelerated and amplified by mistaken macroeconomic policies in the 1980s.[17] America's competitive position began to shift in the 1950s and 1960s in the areas of textiles, footwear, and apparel. These were labor-intensive sectors at the time, and the shift seemed only to indicate a change in the composition of American domestic production; the United States was simply experiencing a natural adjustment of its economic might from labor-intensive to capital-intensive industries. But then from the 1960s through the 1970s, the capital-intensive industries also began to slide. The steel industry provides a good example. Imports of steel rose continuously through the 1960s as competitors from Europe and Asia emerged. As imports rose, American steelmakers responded sluggishly, failing to fully participate in a production revolution from open hearth to basic oxygen furnace technology and from traditional to continuous casting. They never recuperated. By the late 1970s, American steelmakers were not just competing with foreign steel but were having to import the very production technology to do so. By becoming importers of the technology, U.S. steel firms condemned themselves to follow the innovation curve rather than lead it. This same cycle of rising imports, slow domestic response, and relinquishing technological leadership was repeated in other industries. The 1970s then saw the decline of manufacturing and capital equipment sectors—sectors that produced complex assembly consumer durables such as electronics, automobiles, and numerically controlled machine tools. In these sectors, too, American firms missed a production revolution originating abroad. Now, in the 1990s, a range of advanced technology sectors faces intense international competition—including electronic materials and manufacturing equipment, semiconductors, displays and other component technologies, and electronic systems such as computers and office automation equipment. Even high-technology industries are losing market position and feel themselves under siege.[18]

In all of these sectors, there was a characteristic story of retreat culminating in decline. First, global market share began to drop, especially at home as imports flooded the domestic market. American firms usually responded by moving offshore to lower production costs through cheaper labor (if they had the resources) and simultaneously securing some form of bilateral restraint on imports to protect their waning domestic market position. The bilateral agreements established quantitative restrictions on the number of imported items. Placing quantitative restrictions on foreign producers' access to our domestic market encouraged them both to raise prices and to fill their quota by selling their more expensive items instead of less costly ones. This not only raised consumer prices but allowed foreign competitors to capture much of the gain. The policies served only to defer domestic competitive adjustment and even subsidized foreign firms that were becoming ever more competitive. For those who went offshore, the respite was short lived. As production processes became more and more spatially fragmented, integration of product and process innovations became ever more difficult. Finally, as the competitive position of U.S. firms waned, so did their capacity to spend on R&D and new product and process development. The principal sources of innovation and advanced technology development began to move abroad to competitors.

The problems in electronics bring this story of troubled adjustment into the 1990s and demonstrate how deep the loss of U.S. technological leadership runs. Postwar U.S. dominance of the electronics industry was premised on companies' producing complete products, such as computers, while having access to a highly competitive domestic market of independent component, subsystem, equipment, and materials suppliers. Even the most vertically integrated firms (such as IBM) depended on this supply network. Over time, those independent suppliers have been disappearing under competitive pressure from Japanese producers. Today, large, integrated Japanese electronic systems firms control the supply of many of the essential underlying technologies, either directly through ownership or indirectly through group affiliation.[19] This integrated (or group) character of the enterprises is important. Even if components and subsystems are sold on the market, they will almost certainly be made available within the Japanese firm or group first. Indeed, an increasing number of American and European companies, from high-end suppliers such as Unisys and ICL to micro suppliers such as Compaq, Sun, and Apple, sell name-brand computers that consist almost entirely of hardware technologies supplied by their major Japanese competitors.[20] As competitive dependence in the supply base undermines computer product development, the technological initiative is increasingly passing to Japanese industry and they therefore dominate the fastest growing market segments.[21]

This story is repeated in other sectors of the U.S. industrial base. The United States is increasingly dependent on foreign supply of a broad range of industrial technologies including manufacturing machinery, tools and robotics, precision mechanical and magnetic components, displays, optoelectronics, power supplies and control systems, and many advanced materials such as ceramics and ultra-pure silicon. All of these technologies are militarily significant. Advances abroad in all of them have come primarily through civilian markets rather than military spending. Equally significant, access to many of these technologies can be regulated because their supply is controlled by foreign producers with market power. Later we explore the implications of this situation in greater detail. For now, it simply suggests the degree to which U.S. technology development is becoming constrained by decisions abroad—something unprecedented, indeed almost the reverse of the established U.S. position in the postwar period.[22]

From Dominance to Denial

Why don't others see it the way we do? Numerous analyses assert the belief that American industrial position remains fundamentally sound and well prepared for political and market competition in the twenty-first century. In our view, these analyses ignore significant competitive troubles and pay little attention to the long-term constraints on U.S. behavior that those competitive troubles represent. The problems exist in finance, industry, and technological capability, and may be seen from a variety of different vantages.

Consider first the most obvious symbol of the changed global situation: the sudden shift by the United States from its position as the world's largest creditor to that of the world's largest debtor. That shift is also a useful initial lens through which to scrutinize the American economy and its position. For example, there is the issue of what we use debt for: A century ago, when the United States last was a borrower, debt served as an instrument of development that laid the foundation for long-term strength; overseas borrowings were invested in national development. Now, however, they are consumed, leaving only future obligation.

The debt, and the policy of consumption that led to it, has now begun to sharply constrain American fiscal and monetary policy. Macroeconomic management is constrained because substantially eroding the exchange value of the dollar is difficult. As Fred Bergsten argues, such actions could "put into jeopardy the huge capital inflow that remains essential to fund a trade deficit still running at about $100 billion."[23] Such constraints rapidly begin to influence security position. A drooping exchange rate makes the projection of real power more expensive in domestic terms. The dollar cost of overseas operations, everything from United States Information Service (USIS) activities to military bases, goes up. The consumption of foreign product, including elements of security, becomes more expensive.

The U.S. net debtor position will not easily be reversed. Neither Japan nor Europe is ready to volunteer to absorb the massive American exports required over an extended period. Nor has it been simple to accomplish the more modest goal of eliminating the trade deficit, which continues to add to the debt. Certainly with the drop of the dollar, an export boom has begun over the past four years, a boom that has "ranged across virtually the entire spectrum of manufacturing industries. . . ."[24] But just as the deficit and the accumulating debt were not in themselves evidence of radical industrial decline, the export boom is not in itself evidence of a resurgence of industrial competitiveness. Rather, we must look more closely.

This export boom has been sparked at a very low real exchange rate between the dollar and the other principal currencies. American industry can no longer compete at the earlier and higher exchange rates. This amounts to a real shift in the position of the American economy. By 1989, the United States was able to regain its 1972 share of world exports, but only because its manufactures were made extraordinarily cheap by a dollar worth radically less than its 1972 exchange value. Even then, the 1972 trade surplus produced by superior

U.S. export performance had been replaced by 1989's $113 billion deficit.[25] Any country can have balanced trade; the question is, At what real exchange rate and at what real income? The trick is to maintain that trade balance with high and rising real incomes, essentially the definition developed at Berkeley for the President's Commission on Industrial Competitiveness.[26] Unfortunately, average American real weekly earnings in the private sector (nonagriculture) peaked about the time of the first oil crisis and have declined ever since. So we have achieved an export boom by reducing prices (lowering the exchange rate) and reducing real wages. This is, simply, a real change in the national competitive position; our firms can compete, but on different and less attractive terms to the nation.

Shifting the angle a bit, compare the response of American and Japanese producers to currency shocks: The United States once had dominant positions in product and production, partly because it made products others could not make or could not begin to make competitively. Consequently, high American wages and a high U.S. dollar did not displace American producers from markets. Now, however, the situation has changed. The sharp increase in the value of the dollar in the early 1980s priced American goods out of many world markets and made imports a bargain. By the mid-1980s, the United States faced a soaring trade deficit and, with the help of the other major industrial countries, the so called G-7, began to devalue the dollar. But then came the real trouble: As the dollar progressively lost value against the yen, many American industrial producers did not regain their lost market position. In fact, the position of some advanced-technology sectors such as electronics continued to deteriorate. By contrast, Japanese producers of cars and laptop computers retained and even enhanced their market position as the yen rose in value. Japanese producers absorbed the exchange rate–driven price increases of the late 1980s; that is, they had priced to meet the market and accept lower profits.[27] Roughly one-half of the yen's real appreciation was neutralized by a reduction in export prices relative to domestic prices. This did not happen in the United States, where pricing to market—absorbing exchange-driven price increases—was much more limited. Our sectoral observations suggest that determined Japanese firms successfully reorganized their domestic manufacturing operations to increase productivity and flexibly introduce new products as a means of defending market position. In 1987, major Japanese firms announced that they would remain competitive from a Japanese production base even if the yen rose to 120 yen to the dollar. In some segments of electronics, the principal Japanese firms believed they could remain competitive with the yen at 90 to the dollar.[28] We will argue in a moment that such production innovation was possible because Japan is on a distinctive trajectory of production development.

Price elasticities in trade—the sensitivities of imports and exports to changes in currency values—are a related way of considering the real competitive changes. There is substantial debate on whether there has been a change in the trade behavior of American industries.[29] It appears that there has been significant change at least on the import side: During the 1980s, each percentage decrease in the value of the dollar (making American goods cheaper and foreign goods more expensive) resulted in a smaller reduction of imports than it had a decade before; conversely, each percentage increase in the value of the dollar (making American goods more expensive and foreign goods cheaper) produced a greater influx of imports than it had ten years prior. That is to say, for example, that if in 1977 a 10 percent devaluation of the dollar reduced imports by 7 percent, in 1988 that same devaluation reduced imports by only 4 percent; and if in 1977 a 10 percent rise in the dollar led to a 5 percent increase in imports, in 1988 that same rise gave way to a 9 percent increase in imports.

This trend is consistent with the observation that a substantial increase in American imports has been concentrated in automobiles and electronics over the past decade. Much of the deficit in these sectors is in products American producers do not make or undeniably do not make as well as the Japanese, so we would expect such imports to be less sensitive than formerly to price changes. That real sectoral shift in competitive position represents a change in the industrial structure of the economy. It is consistent with econometric evidence that a given rise in domestic prices produces a larger inflow of imports than when American producers made a distinctive basket of products or made the products distinctively well. By contrast, the same evidence suggests that there was not a change in the price elasticity of exports. This appears to be an artifact of the composition of U.S. exports: What U.S. companies continue to sell well abroad (e.g., agriculture, chemicals, aircraft) is as sensitive or insensitive as they have always been to price changes.

A final way we might judge the relative decline of the U.S. position is to consider the troublesome composition of our trade deficit. The mix of imports that our deficit embodies is evidence of changed industrial position. If American auto producers and consumer electronic producers were highly competitive with their Japanese counterparts, then we would import fewer cars and VCRs. We might, rather, import all of the French wine harvest or traditional Japanese artifacts and crafts. Recent studies of advanced countries' trade patterns have recategorized all of the sector-level data to examine this problem.[30] They show a radical loss of U.S. position in traditional and scale-intensive industries such as textiles and consumer durables, and in production equipment (capital goods) and materials sectors, where, for two generations, American producers have dominated. These latter sectors in particular embed substantial industrial know-how and, as we argue below, provide an important foundation for future growth. By contrast, Germany and Japan have maintained or gained position in traditional, scale-intensive, and production equipment/materials industries. Moreover, relative to the United States, Japan has claimed position in advanced-technology sectors. The relative inability of American producers in diverse sectors to compete in global markets has shaped not only the composition of our trade deficit but also our industrial base.

The perspectives just examined above suggest a very different story about American industrial development than the conventional tale that was presented in the early 1980s. At that time, a positive face was placed on trade deficits in older industries: The deficits were supposed to represent a shift upward out of declining labor-intensive into expanding technology-intensive industries, a shift out of "sunset" into "sunrise" industries.[31] The apparent decline in these supposedly mature sectors was claimed to be a source of strength for the economy as a whole. Unfortunately, the sunrise–sunset distinction simply misinterpreted many of the processes of industrial development. High-tech sunrise sectors largely make producer goods: equipment, components, subsystems, machinery, and advanced materials used to produce or develop final products for consumers. The sunrise-sector goods are applied across the economy to help transform production and products in traditional industries. The traditional industries are vital clients. Without demand from the traditional sectors for high-tech products, the domestic component, systems, and equipment producers cannot develop.[32] Conversely, this supply base of component and equipment producers embodies vital skills and knowledge to help sustain production and product innovation in their clients. This interlinked character of industry, this industrial fabric, matters.[33] Rather than cause for optimism, the shift out of sunset industries boosts real elements of concern: The U.S. inability to maintain position in "mature" sectors, conjoined with the success of Germany and Japan in the same sectors, suggests that U.S. firms have a limited ability to reorganize manufacturing and apply new technology. This simultaneously weakens the advanced-technology sectors of the domestic economy by eroding their customer base.[34]

Our interpretation that the late 1980s resurgence of American industrial production does not indicate a return to an equivalent competitive position is supported by recent comparisons of the advanced industrial economies.[35] The United States is certainly one of the wealthiest countries in the world. But it is revealing to decompose the measure and source of its wealth, and to compare those with the sources of wealth of the other twenty-five richest countries. First, consider gross national product (GNP) per resident. Here the United States ranks second. However, in gross *domestic* product (GDP) per resident— which excludes imports and exports—the United States ranks only fourth. The gross domestic product in manufacturing per inhabitant tells a worse story. The United States ranks eighth, behind France and barely ahead of Denmark. This position is a radically new development. In 1965, the same measure placed the United States in the first position. By 1973, the United States had fallen to third position. In 1981, the United States had fallen to fourth position and by 1984 to tenth—a total of nine places between 1965 and 1984; America's current resurgence regains only two of those places.

Perhaps this erosion of U.S. manufacturing position does not matter. Can't the service sector substitute for manufacturing? This issue has been examined in detail by Stephen Cohen and John Zysman.[36] Our position is that services cannot substitute for manufacturing as a means of supporting either the relative domestic standard of living or the U.S. international trade position. Critical areas of the service sector are linked to manufacturing, and their capacity to support income growth will erode as manufacturing loses position. Indeed, the drops in real wages examined earlier in part reflect the exit from manufacturing to services employment. Service industries will also not compensate for manufacturing's trade deficit. Not only are internationally traded services a small fraction of total trade, but the U.S. position in services is weakening, as the relative position of American banks and other financial intermediaries suggests. The U.S. competitive position is not likely to find solace in the myth of the postindustrial service economy.

Neither can comfort be found in the observation that the American share of global GDP, after declining until the mid-1970s, has since remained stable.[37] The GDP figure is very deceptive. The continuing U.S. position rests not on excellent performance, but on an economic slowdown in Europe in the 1980s. By contrast, growth rates in Asia have remained higher than in the United States; the U.S. position relative to Asia and Japan has continued to decline. For security purposes, that relative decline matters. That the United States has been able to "hold its own" with a temporarily sclerotic Europe is hardly reassuring. A European resurgence without a parallel American boom would diminish the U.S. position further.

Finally, can "soft power" substitute for a decline in industrial position? American ideas, values, and culture have been widely admired and have almost certainly extended U.S. influence. More importantly, they have often given legitimacy to our power. But the role of soft power should not be exaggerated: it is a weak hand to play in an era of decline. Espousals of democracy cannot, for example, compensate for delays in providing real resources to aid the transformation of the Soviet Empire. The spread of American culture may extend our influence for a time, but it has been the economic wealth of the United States that has transmitted our culture in the first place. As our economic position declines, so certainly will the influence of our soft power. The culture of a declining power does not command acquiescence. Japan is now the model explored in the endless airport paperbacks; Germany is the centrally debated alternative for a France looking for guidelines for its own future.[38] Unbacked by real influence from economic power, our culture, our "soft power," may even eventually elicit contempt or dismissal—the weapon of a paper tiger. Debates in Japan and Europe about weakness in our society suggest that may already have begun.

The Political Meaning of Economic Decline

What is the political meaning of these economic changes in U.S. finance, trade, and technology positions? Stripped of political context, these economic statistics understate the loss of political position. They cannot capture the political significance of European integration, German unification, or the rise of Japan. These are new, autonomous players whose emergence challenges the old constructs of American influence. To appreciate the relative decline of the U.S.'s political position, we can compare the percentage of American GDP to the GDP of its two largest autonomous competitors over the past two decades.[39]

In 1970, the two largest fully autonomous western competitors to the

United States were Britain and France. In that year, American GDP was 3.7 times that of the two combined.[40] In 1970, the American GDP was also 2.5 times the combined GDP of Japan and Germany—larger economies, but not yet politically autonomous. By 1990, Japan and Germany had become autonomous political players, and the relative size of the combined economies had roughly doubled. In 1987, the American GDP was only 1.3 times that of Germany and Japan combined. The 1987 indicator even understates the change in the balance of power, because since 1987, differential growth rates have continued and the two Germanies have become one. Finally, of course, if the European Community (EC) is considered a single actor for purposes of this comparison, the situation would be even less favorable to the United States.

America's economic position relative to its two strongest allies has gone from almost 4 to 1 in 1970 to virtual parity by 1990. To be sure, translating economic resources into political influence is another matter, but it is clear that the relative economic position of the United States has changed sharply. Political consequences are surely inevitable. As we learned from the Gulf War, the American capacity to extract compliance from its allies in the security system has already diminished. Politics among nations is only beginning to reflect this real change in capabilities. Consequently, recent history is no guide to the rest of the decade.

II. Innovation Abroad and Constraint at Home

The American position in global manufacturing competition has changed abruptly.[41] After World War II, the United States made things others could not produce, and what others could make, American firms often made better and more cheaply. America's dominant industrial position rested on a system of mass production and divisionalized management that emerged in the late nineteenth and early twentieth centuries, and that was strongly supported by domestic policies favoring consumption.[42] These real innovations in the organization of production and corporate control were responses to the particular circumstances of American economic development.[43] Other countries tried to catch up. They sought to imitate what we did; they saved and invested to do so. But they never really did imitate the United States. Rather, the most successful innovated and built the basis for advantage in global markets.

Two aspects of postwar development in the foreign advanced countries concern us: policy and production. The two stories intermesh. Consider first the policy questions. Our most successful competitors, such as Japan and Germany, chose to emphasize investment in production over consumption, creating macro conditions for rapid growth. In both cases, governments encouraged the rapid adoption and widespread diffusion of technology acquired abroad, and helped provide for a corresponding skilled workforce. In Japan, the government went a step further. Not only did government stimulate new investment through a variety of tax incentives, but by formally closing the domestic market to foreign firms, it reserved growth in domestic demand for Japanese producers. As technology followers, Japanese firms borrowed, implemented, and improved foreign technologies through continuous rounds of reinvestment in the rapidly growing domestic economy. In essence, Japanese firms faced conditions in traditional industries that Americans associate with high-technology industries—rapid growth and technological development forcing dynamic adaptation through investment and learning. Learning economies dominated, making the pursuit of market share a necessity to sustain short-term profits.[44] Continuous rounds of new production investment, supported by rapid growth policies, helped to create a virtuous cycle of productivity gains. Each new round provided an opportunity to experiment with production, achieve new scale economies, adopt and refine new technologies, and build an iterative pattern of learning while doing so. Real innovations in production and in technology development were generated and accumulated. Japanese approaches to institutional structure and economic policy thus created patterns of market logic (and subsequent corporate strategies) distinct from other nations.

Accompanying these new government approaches to rapid growth come innovations in production and production organization in countries as diverse as Japan, Germany, and Italy. Our hypothesis is that these breakthroughs are of sufficient scope and power to alter the relative position of nations.[45] What is emerging is not incremental or even radical improvement of an old system, but a new approach, a new paradigm. Elements of these breakthroughs are found in the U.S. industries, but the evidence is that the new approach is not as well established or as diffused in this country as it is elsewhere.[46]

The detailed character of the production revolution is increasingly understood and documented.[47] The central code words of the new manufacturing are *flexibility, speed,* and *quality.* The popular notions of quality circles, just-in-time delivery, and automation—slogans of the new approach—are simply organizational or technological elements of the whole. The flexibility of the new manufacturing, rather than these slogans, better signifies the revolution in production. Manufacturing flexibility consists of two important capabilities: static flexibility and dynamic flexibility. Static flexibility is the capacity to vary product mix on a single production line or to automate batch production; dynamic flexibility is the capacity to introduce new production methods and new products without significant disruptions to existing set-ups and practices.

The organizational and technological innovations that permit flexibility have actually been implemented in a variety of forms. One form that has attracted considerable attention is so-called flexible specialization.[48] Popular in northern Italy and parts of Germany, this model involves an attack by smaller firms on niche markets. It is built on craft skills and on local community infrastructures that permit shifting ties between firms that compete one day and collaborate the next.

By contrast, the most powerful implementation of the new manufacturing involves flexible volume production (labeled variously as flexible automation, flexible mass production, and lean production).[49] Until recently, high-volume production has been dominated by the rigidities of scale economies: expensive equipment dedicated to specific tasks in which the costs could only be recouped by large production runs of the same items. Introducing variety was very costly because it disrupted production runs and incurred significant costs by requiring long set-ups and substantial down time. Now, organizational innovation, reinforced by microelectronics, has removed past constraints. The new approach creates the capability of producing a variety of tailored products with costs, quality, and market responsiveness far superior to mass production.

Japanese firms have been the most successful at implementing this new production system by creating, as Jon Krafcik has noted, a relatively "lean" manufacturing process that tends to use less capital, fewer people, and produce less waste than traditional mass production.[50] The lean production approach is characterized by shorter production runs manned by smaller teams of multi-skilled workers operating less expensive general-purpose machinery that can be rapidly changed over for new production set-up with minimal down time.[51] Line workers are given responsibility for strict process control in order to eliminate (systematically) variability in manufacture (the major source of defects). In turn, elimination of defects and rapid changeovers eliminate the need for carrying inventory and permit parts to be delivered as needed—"just in time" for production—further reducing costly inventories. Tight process control and the multi-skilled work team also eliminate the costly layers of supervisory, maintenance, housekeeping, and quality control personnel that characterize mass production. Significant gains in product quality without increased costs are one landmark result of the overall system. Even when very long production runs in the tradition of mass production are used (as in the manufacture of common underlying components such as semiconductor random access memory) the result is higher yields and lower costs due to better equipment utilization and superior control of the sources of defects.[52] The truly remarkable competitive power, however, comes from the inherent flexibility that shorter production runs provide. Indeed, the new manufacturing system is designed to accommodate change in both the static and dynamic senses identified earlier (i.e., varying product mix and accommodating new production methods that permit wholly new products).[53]

The new flexible manufacturing system actually extends beyond the shop floor into product development and to suppliers. Tight links between design, development, and manufacture permit "design for manufacturability"—with minimized parts counts and mutual accommodation between product and process requirements—resulting in the lowest possible anticipated costs and fastest development cycle times. Tight links to skilled suppliers, often structured through partial ownership and long-term business relationships, permit suppliers early involvement in product definition, aim at assessing and reducing overall costs, and permit a fair allocation of costs and returns between suppliers and final producers. Finally, the whole system lends itself to automation without becoming rigid. The overall result is great flexibility in production and greatly reduced total cycle times, thus enabling superior market responsiveness. Indeed, the flexible, speedy production capability permits the leading firms to do their market research by introducing a new product and then adjusting to customer reaction, fine-tuning product configurations and volumes to actual demand.[54]

Many aspects of this new production model—for example, the changes in accounting practice required to express management choices in terms of speed and capital productivity rather than labor costs—are still being worked out and remain to be fully described and theoretically supported. But the new practices are already transforming traditional industries—generating vertical disintegration in many cases, new entry in other cases, and prying open established industrial structures.[55] The new forms of production suggest a sharp break from practices dominant in the middle part of this century and pave the way for realizing the huge gains in productivity that have been promised but not yet delivered by the application of information technology to production.[56]

That potential is strongly underlined in the remarkable work of Ramchandran Jaikumar, depicting the historical evolution of the technology and management of process control.[57] He argues that manufacturing has evolved through six stages, each representing a change in how people thought about and practiced manufacturing. At each step, manufacturers addressed and ultimately controlled different major sources of variance in production, leading to order-of-magnitude advances in productivity and quality (using product rework as a measure of quality). Jaikumar's first three stages are: (1) the original emergence of machine tools in England, (2) the establishment of the American system with measurement, special-purpose tools, and interchangeable parts, and (3) the Taylorist system of managing the time and motion of labor. Each stage reflected an increase in production scale, increased specification of tasks before production began, and more rigid control of the system once in operation. That rigidity meant the system was quite static, capable of only limited response to the unexpected inside or outside the production system. Taken together, these stages culminated in the post–World War II American system of mass production.

By contrast, Jaikumar's next two stages introduce a much more dynamic manufacturing capability and are intimately related to the new production innovations described before. Those stages are (4) the introduction of statistical process control and other means of identifying and systematically controlling the sources of variance, and (5) the introduction of automation through information processing and numerical control. These factors create more dynamic and adaptive capacities because they require increasingly detailed specification of the production process and enable anticipation and response rather than rigidity in production. Combined with the accompanying organizational innovations described earlier, these stages culminate in the current model of lean and flexible production.

The sixth step, on the horizon with no leader yet established (and perhaps somewhat beyond current technical and organizational capacities), is the emergence of truly intelligent and fully integrated systems of computer-controlled manufacturing (i.e., computer-integrated manufacturing—CIM). Taken together, the last three stages promise near-real-time adaptation to market changes with extraordinary levels of flexibility, productivity, and quality. In short, when new information technologies are added to the dramatic changes in organization and management, the current transformation in manufacturing practices may well have the potential to generate discontinuous leaps in performance and innovation. If the past is any guide—and Jaikumar's work suggests that it is—such a discontinuous jump in production capability will create distinct competitive advantages for firms and nations that master the new system. As we argue in the next section, it could place them on a more rapid growth path at least until the new capabilities fully diffuse to others.

In our view, America's relative decline reflects a failure to understand, access, and adopt the innovations in policy and production that underlie superior industrial performance abroad, especially in Asia. Myth and flawed practices have also deterred the competitive responses of U.S. industry.[58] There has been an unwillingness to acknowledge that fundamentally different practices are at the root of the competitive problems. Many executives believed that technological leadership could be maintained indefinitely even as manufacturing mastery was ceded to competitors, or that foreign labor or cheap capital costs rather than production innovation lay behind superior performance abroad, or even that refined techniques for financial management made long-term strategic planning, production reorganization, and technology investment unnecessary.[59] Many U.S. firms have finally begun to overcome these myths. They are slowly undertaking a strategic reconceptualization of the firm and its place in the market and the community—the necessary prerequisite to adopting new production innovations.[60]

Although the glimmer of change is hopeful, the decline in industrial position and relative incapacity to adjust to the new production model will be difficult to reverse for several reasons. First, although some firms are changing, many more have not even begun to reconsider their practices and strategies. This is particularly true for the bulk of small and medium-sized U.S. manufacturers throughout the domestic industrial base, as foreign manufacturers doing business with them affirm.[61] Second, even those U.S. firms making the necessary changes have discovered that, once displaced from markets during the period of the high dollar, they no longer have product or cost advantages that permit them to recapture their lost position. Third, as the next section argues, the altered industrial position is leading to dangerous dependences that constrain the ability of U.S. firms to adopt the new production model.

Diffusion and Adjustment: The Risk of Competitive Dependence

Far more has been lost than simple market position in specific sectors. Rather, the supply base of the economy is unraveling: The components and parts technologies, materials and machinery sectors, and related industrial skills necessary to sustain competitive manufacturing and development are eroding, or are already gone.[62] For example, competition in the past decade has devastated domestic producers of manufacturing machinery, including advanced industry segments such as computer-numerically controlled machine tools, robotics, and semiconductor photolithographic equipment. U.S. dependence on foreign supply of such machinery has increased dramatically since 1988, with imports rising from 14 to 40 percent of domestic consumption.[63]

Table 1-1 similarly shows that, in electronics, U.S. producers are broadly dependent on foreign supply of a huge and growing list of essential component, materials, and machinery technologies. Indeed, most U.S. computer firms can no longer produce consumer-like products (e.g., laptop and smaller PCs) without an alliance with Japanese firms to provide the necessary components, micro-design know-how, and relevant manufacturing skills—Compaq with Citizen Watch, Apple with Sony, Sun with Fujitsu and Toshiba, and Texas Instruments with Sharp. Even IBM is not immune from this trend. The U.S. General Services Administration recently noted that IBM's RISC System 6000 model 7013–540 computer has a foreign content in excess of 88 percent.[64]

Of course, the significance of this competitive dependence is open to debate, particularly where required technologies are readily available from

Table 1-1 Gaps in the U.S. Technology Supply Base

Precision-mechanical
- Motors—flat, high torque, sub-miniature
- Gears—sub-miniature, precision machining
- Switch assemblies—sub-miniature

Packaging
- Surface mount, plastic

Media
- Magnetic disk
- Optical disk

Displays
- Electroluminescent
- LCD, color LCD, LCD shutter
- CRT—large, square, flat
- LED—arrays
- Projection systems

Optical
- Lens
- Scanners
- Laser diodes

Feromagnetic
- Video heads
- Audio heads
- Miniature transformer cores

Copier-printer
- Small engines for laser printers

Source: National Advisory Committee on Semiconductors

abroad. In industries such as textile production, for example, several U.S. producers (e.g., Milliken) have been able to remain competitive through rapid adoption of machinery imported from a variety of competitive European sources. Other American industries have fared less well, however. In consumer electronics, U.S. producers became competitively dependent, lost the capacity to keep pace with product and process innovations occurring abroad, and, eventually exited the market almost entirely. Similarly, U.S. automobile producers face a weak domestic parts and components supply combined with difficulty in accessing supplier innovations abroad.[65]

In electronics, existing dependencies appear to be slowly creating a cumulative knowledge gap that is profoundly disturbing in its security implications: Even when they can procure technology inputs from abroad, U.S. firms no longer retain many of the design and manufacturing skills necessary to use them in a competitive fashion. For example, Japanese producers have painstakingly acquired, iteratively over several product generations, the precision mechanical design expertise embedded in products such as VCRs, or the precision machining know-how in auto-focus camcorders. A leading U.S. industrial laboratory recently reverse-engineered such products and concluded that the embedded precision mechanical skills probably no longer existed anywhere in the United States.[66]

These are serious challenges to America's security position because components, materials and equipment manufacturers increasingly control the technological advances in product and production know-how that help to shape competitive performance. Competitive dependence will increasingly constrain the adjustment of U.S. producers by deterring access to appropriate technologies in a timely fashion at a reasonable price.

The Architecture of Supply and the Trajectory of Technology

It is not the fact of dependence on foreign producers itself that concerns us. It is rather the "architecture of supply"—the structure of the markets through which components, materials, and equipment technologies reach U.S. producers.[67] Again, by the supply base of an economy, we mean the parts, components, subsystems, materials, and equipment technologies available for new product and process development, as well as the structure of relations among the firms that supply and use these elements.[68] The supply base can be thought of as an infrastructure to any given firm, in the sense that it is external to the firm but broadly supports the firm's competitive position by helping to delimit the range of its possibilities in global markets, while providing collective gains (e.g., technological spillovers) for the economy as a whole.[69]

The supply base affects producers by enabling or deterring access to appropriate technologies in a timely fashion at a reasonable price. The architecture (or structure) of the supply base matters to the extent that it influences such technology access, timeliness, and cost. Domestic industry that is significantly dependent on a foreign supply base (i.e., on imports of key inputs) will not be overly constrained wherever markets are open and competitive, and foreign suppliers are numerous, geographically dispersed, and not in the same lines of business as their customers. This was essentially the case for European electronics systems producers from the 1950s to the 1980s: They relied primarily on U.S. components suppliers, who were themselves competitive, numerous, located in both Europe and the United States, usually not in competition with their customers, and accessible through relatively open markets for trade and investment. Indeed, it was not until the competitive problems of U.S. chip producers threatened a much more constraining architecture of supply for Europe in the 1980s that European companies moved at great cost to re-create a locally controlled supply base.[70]

By contrast, domestic producers should be concerned where the architecture of supply is characterized by closed markets, oligopolistic and geographic concentration, and, especially, wherever such concentrated suppliers compete directly with their customers. When suppliers have the ability to exercise market power or to act in concert to control technology flows, or when markets and technologies are not accessible because of trade protection, then the architecture of supply can significantly constrain competitive adjustment to the disadvantage of domestic industry. Such an architecture is emerging today in American electronics production: A small number of foreign suppliers, principally Japanese, are more and more driving the development, costs, quality, and manufacture of the technological inputs critical to all manufacturers. Most of these suppliers of electronic components, manufacturing equipment, and subsystems are also competitors in a range of electronics systems from TVs and portable phones to computers. These competitors are then increasingly in a position to dictate the degree of access U.S. producers have to essential technologies, the speed at which they can bring new products incorporating them to market, and the price they pay for the privilege.[71]

The supply base architecture thus becomes a crucial element of international competition for domestic industries. It has an even greater significance, however, for the domestic economy. The architecture of supply and the composition of domestic production together delimit the technological opportunities that are perceived and pursued within a domestic economy.[72] They define a technological development trajectory that reflects the community and market context within which technology evolves.[73]

Such development paths are not dictated by technical knowledge alone. Historical studies of technical change suggest that technological advance is open-ended rather than preset by scientific blueprints. Development, production, and use—and the learning they entail—shape the evolution of technologies at least as much as does scientific research. Technology is a path-dependent process of learning in which tomorrow's opportunities grow out of product, process, and applications activities undertaken today.[74] Consequently, the pace and direction of technological innovation and diffusion are shaped by production and market position.

In this view, technological know-how cannot simply be acquired through international market mechanisms; otherwise, there would be no possibility for distinctive national development trajectories. To be sure, some technical knowledge is purchasable in disembodied form, such as a blueprint or a dress pattern. Even more know-how is embodied in products and can be accessed through purchase and elaborated through use. But much technical knowledge involves additional, often more subtle insights that coalesce only in conjunction with experience in development and production. The process is simultaneously cyclical and incremental—rather than a dramatic leap up to the next rung in the ladder of technological progress, advances are driven through iteration and cumulative learning by doing in production.[75]

This kind of technological knowledge differs considerably from pure science and is supported by different practices and institutions. Scientific knowledge, with its theories, principles, and premises, often can be precisely specified and easily communicated in a common language. Western institutions of science boast a history of openness, are international in scope, and permit information to flow readily across national borders. By contrast, the technological knowledge generated in production and development usually accrues locally and, under the right circumstances, can be kept from diffusing for considerable periods of time.

As local learning occurs, such know-how accumulates in firms as a skilled workforce and proprietary technology and techniques—all of which are usually difficult to copy because they have been painstakingly acquired through iteration over time. Such technological know-how also accumulates in local communities in the production networks of suppliers and contractors, and in social networks among technical peers. It can also amass nationally in the cumulative skills and experiences of the workforce and in relevant national institutions (e.g., national laboratories, universities, or specialized agencies that diffuse technology, such as the U.S. Agricultural Extension Service).

The speed and degree to which such embodied technical know-how flows across national boundaries depends crucially upon the character of these local and national institutions. In the United States, labor mobility is very high, firms can be purchased outright, and short-term capital market constraints often push firms to license proprietary technologies. Social and production networks are relatively open and fluid, and many relevant national institutions are accessible (e.g., universities and national labs). In general, U.S. technology accrues locally but diffuses rapidly even across national boundaries. By contrast, in a country such as Japan, skilled labor mobility is low, acquisitions are virtually impossible, patient capital (i.e., capital willing to wait for returns) is available, and relevant networks and national institutions are extremely difficult to access. As a result, considerable accrued technological know-how is retained locally in Japan and never diffuses readily or rapidly across national boundaries.

Because know-how can accumulate and be retained locally, the character of the domestic economy and the architecture of the supply base supporting it can dramatically shape the availability of national technological opportunities. It is our hypothesis (which we elaborate more fully in Section IV) that three regional supply architectures will emerge in Asia, America, and Europe. The structure of each—the mix of skills, components, subsystems, equipment, and technological ideas—will powerfully affect the terms on which international competition evolves. Rather than global markets displacing national economic foundations, we see regional structuring.

Those regional architectures and the technology development trajectories they support will influence the speed and extent of adoption of the new model of lean, flexible production, and thus of competitive adjustment in each region. In our view, the American architecture of supply is increasingly limiting. Capacities to access and adopt the new production model are severely constrained, as is access to the technologies essential to the new model's future evolution.

Competitive Dependence and Defense Capabilities

There are real implications for America's defense position. Boldly put, why can't we simply depend on foreign suppliers in an open international economy? Theodore H. Moran, for example, contends that the threat of foreign control is a function of how few or how many firms are relied upon in a given defense industry, but not of the nationality of firms per se. Therefore, according to Moran, the most dependable method for minimizing the threat of foreign control is simply to diversify and multiply the companies upon which a nation can draw for its technology base. In Moran's construct, the potential for foreign control decreases in direct proportion to the proliferation of suppliers (irrespective of their nationality). As we have implied by contrast,[76] corporate nationality can still matter a lot and policies aimed at the proliferation of suppliers are an inadequate response in cases where a small number of firms, all located in one country, already dominate a world industry. Corporate nationality matters if it bears at least some relationship to influence and control. As our colleague Stephen Cohen maintains, "We do not yet live in the age of the 'global corporation' nor, in its logical concomitant, a world of politically undifferentiated spaces. We should not assume that all multinational corporations [MNCs] are the same. All Home countries do not treat their MNCs the same; and all Host countries do not de facto set the same conditions for behavior on all MNCs."[77]

We agree with Moran that ownership should not be the defining consideration for U.S. policy; behavior should. Behavior reflects both ownership and residence. It is corporate behavior—what companies do and don't do within a country and with that country's people—that directly determines the wealth and power of that country. Ownership, as we learned during Wall Street's recent takeover binge, still influences corporate behavior. As Moran's own work demonstrates, even the most global of multinationals will take orders from their home governments when circumstances are exceptional. The reaction of American-based MNCs to Reagan Administration entreaties against the proposed Soviet–European gas pipeline is an instructive example.

As we argued in the previous section, domestic industry that is significantly dependent on a foreign supply base will not be overly constrained wherever markets are open and competitive, and foreign suppliers are numerous, geographically dispersed, and not in the same lines of business as their customers. Where we differ with Moran, is in the degree to which one can presume an international economy already characterized by substantial globalization and interdependence. That is one vision, and a desirable one, especially given the security order it would imply, but it is not the only possible future. The United States response to foreign competition and foreign direct investment should certainly include policies aimed at creating the conditions of an ideal global, interdependent world—where strategic industries are structured by a large number of companies, located in a large number of countries. But economic security policy cannot be confined to encouraging a proliferation of suppliers in the face of a very different, preexisting competitive dynamic; that is, in cases where a small number of firms, all located in one country, *already* dominate a world industry. This, we have just shown, is occurring in electronics.

Legislating absolute and universal rules to deal with situations that are so far from universal and absolute seems to us to be the wrong approach.[78] Domestic producers need to be concerned when the international supply base on which they depend for critical inputs is characterized by closed markets, oligopolistic and geographic concentration, and especially when such concentrated suppliers compete directly with their customers. When suppliers have the ability to exercise market power or to act in concert to control technology flows, or when markets and technologies are not accessible because of trade protection, a universal rule will not do. The United States needs to promote a diverse and open international market in advanced technologies. But the United States will not be able to access this international supply base without having resident in the US the essential skills and production capabilities necessary to apply technologies available abroad. Assuring the domestic supply base is the only means of achieving the goal of openness and interdependence.

The innovations abroad in production and policy that have helped to create competitive dependence are simultaneously providing a new and different technological foundation for the security system. As the next section argues, not even America's superior defense technology capabilities are likely to rescue the United States from this dilemma.

III. Spin-Off versus Spin-On: Old and New Defense Technology Trajectories

The industrial economy is eroding at precisely the wrong time for America's security concerns. American military technology will not rescue the commercial U.S. position. Rather, a weakening commercial position will almost surely affect U.S. capacities to develop military technology and systems. The links between military and commercial technologies are shifting, and so is the relative contribution of each to security. Consequently, national military capabili-

ties must be reassessed and their compass redefined. That reassessment will not
be easy because it will depend upon the always contingent details of technology
development at a given historical moment.

The relations between the civilian and military industrial sectors change
over time. Commercial factors have always influenced the technological opportunities available to support the U.S. security position, even as military spending has shaped the civilian economy's composition and character. The relative
contribution of each to security and the domestic economy shifts; the movement is closer to a pendulum than a progression. Early on, development of
mass production and interchangeable parts was accelerated by military demand
for rifles during the Civil War. The two World Wars saw the organizational
and technological innovations of commercial mass production establish America's ability to churn out huge numbers of tanks, guns, and planes. In those
days, the defense production base grew directly from the commercial production base. The precursor to a new model of technology development was taking
shape, however, as directed government spending created new defense technologies including radar, artificial rubber, the atomic bomb, and the rocket.

The new model fully took hold in the United States after World War II,
helping to create a new technological development trajectory. The model was
premised on the belief that putting investment into science at the front end of
the development pipelines would produce technology at the other end. Military
and related spending (e.g., space exploration) supported the enormous development costs of relevant new technologies. Initial applications were developed for
(and procured by) the military, and later would diffuse—"spin-off"—into commercial use. In this way, U.S. defense spending promoted the rapid development of jet aircraft and engines, microelectronics, computers, complex machine
tools, advanced ceramic and composite materials, data networks, and a host of
other relevant technologies.

Very often, the model worked well to establish both defense and civilian
technology leadership. In the jet aircraft and semiconductor industries, for
example, government priorities helped to set the functional characteristics of
the emerging technologies, R&D funds accelerated the development of the
technology, and military procurement at premium prices constituted a highly
effective initial launch market.[79] A variety of mechanisms, ranging from patent
pooling to loan guarantees for building production facilities, helped to lower
entry costs, diffused technology widely among competitors, and set the stage
for commercial market penetration. U.S. defense policy thereby helped to create advantage and foster competition in the later style of Japan's Ministry of
International Trade and Industry (MITI).

That kind of technology development trajectory continues in some
instances to be successful for the United States. Recent commercial spin-offs
from military spending include local area networking, gallium-arsenide components, massively parallel computing, and algorithms for data compression.
But even in the heyday of U.S. technological leadership, this development
model had occasional problems in transferring technologies from defense to

civilian markets in a timely and competitive fashion.[80] For example, the U.S. Air Force supported the development of numerical control technology for machine tools building advanced aircraft. The programming language proved too complex for general commercial use. Diffusion was slow and civilian application costly. In this case, the spin-off trajectory produced only a commercially vulnerable U.S. industry that was squeezed by Japanese competitors from the low end and German firms from the high end.[81] The foreign competitors benefited from different development trajectories. The Japanese built around MITI support and low-cost, simple technology for general-purpose commercial applications. The German craft tradition of high-quality capital goods, and its institutional support in local government, trade associations, and universities, fostered cost-effective numerical control (NC) technology for high-precision commercial uses.

As the NC experience demonstrates, there are competing technology development models. Massive resources committed in specialized defense contractors to technology produced in batch processes for initial use in military projects constitutes one development trajectory. Massive resources committed to commercial development produced in volume for consumer markets constitutes a separate trajectory. The former is the development model that has underpinned U.S. leadership of the postwar security system. The latter has underwritten the increasing Japanese success in commercial markets. The problem for the security system is that the latter trajectory is proving to have increasing military relevance.

From Spin-Off to Spin-On

A completely alternative military technology development trajectory is emerging from the innovations in production and consequent reshuffling of markets examined earlier. This alternative drives technological advance from commercial rather than military applications. Technology diffuses from civilian to defense use rather than vice versa, a trajectory characterized as "spin-on" in contrast to its predecessor. The new alternative is prospering most fully in Japan, where an increasing range of commercially developed technologies are directly, or with minor modification, finding their way into advanced military systems.[82] In particular, militarily relevant sub-system, component, machinery, and materials technologies are increasingly driven by high-volume commercial applications that produce leading-edge sophistication, with extremely high reliability but remarkably low costs.

The case is clearest in electronics, where a new industry segment is being defined in Asia, largely outside of U.S. control and with only limited U.S. participation. Its distinguishing characteristic is the manufacture of products containing sophisticated, industrially significant technologies, in volumes and at costs traditionally associated with consumer demand. Such products include the latest consumer items, such as camcorders, electronic still cameras, compact disc players, and hand-held TVs, and new micro-systems, such as portable faxes, copiers and printers, electronic datebooks, laptop computers, optical disk mass storage systems, smartcards, and portable telephones. This "high-volume" electronics industry is beginning to drive the development, costs, quality, and manufacture of technological inputs critical to computing, communications, the military, and industrial electronics. At stake is a breathtaking range of essential technologies from semiconductors and storage devices to packaging, optics, and interfaces.

Such products contain, for example, a wealth of silicon chip technology, ranging from memory and microprocessors to charge-coupled devices (CCDs), and have been a principal factor behind the drive for Japanese semiconductor dominance. Over the past decade, emerging high-volume digital products have grown from 5 percent to over 45 percent of Japanese electronics production, accounting for virtually all of the growth in domestic Japanese consumption of integrated circuits (ICs).[83] With this segment continuing to expand at 22–24 percent per year, more than twice as fast as the approximate 10 percent per year average growth rate of the electronics industry as a whole, high-volume electronics will constitute an ever-larger part of the electronics industry of the next century. Its impact on the component technologies that military systems share is just beginning to be felt.[84]

Aside from silicon integrated circuits, militarily relevant optoelectronic components such as laser diodes and detectors, LCD shutters, scanners, and filters are also present in the new high-volume products. For example, the semiconductor lasers that, at different wavelengths, will become the heart of military optical communications systems, are currently produced in volumes of millions per month, largely for compact disk applications. Displays and other computer-interface technologies provide yet another significant overlap of high-volume and military markets. Miniature TVs from Japan are the leading users of flat-panel, active matrix, liquid crystal display (LCD) technology—a technology that is just beginning to infiltrate military systems. Map navigation systems beginning to appear in automobiles are the functional equivalent of military digital map generators.

Optical storage was refined for consumer compact and laser discs, but is likewise beginning to spread into military applications, as are the latest miniature commercial power technologies (e.g., batteries for portable phones). High-volume requirements are also driving a wealth of imaginative packaging technologies that range from tape automated bonding and chip-on-board to multichip modules. Producers of hand-held LCD TVs already use packaging technology as sophisticated as that being used in advanced U.S. defense systems. The new electronics products are driving similar innovations in precision mechanical and feromagnetic components such as motors, gear and switch assemblies, and recording heads, transformers, and magnets. Ball bearings used in videocameras, for example, are now of equal precision to those required for missile guidance systems.

Successful production for high-volume markets also requires mastery of several different kinds of highly responsive product development, materials, and manufacturing skills. For example, Japanese consumer producers, such as

Matsushita, now supply the most advanced manufacturing equipment for IC board insertion, a capability essential for military systems production. Similarly, because elaborate repair and maintenance are not cost-effective in consumer markets, high-volume producers deliver product reliability levels that often now surpass military products at far less cost. The most advanced high-volume electronics suppliers, as we have noted, do their market research by introducing products and fine-tuning product configurations and volumes to meet actual demand.[85] They are the masters of the new manufacturing—utilizing an extremely short and efficient development cycle, and flexible, low-inventory manufacturing.

In sum, the basic technological requirements of new consumer products now approach, equal, or at times surpass those needed for sophisticated military applications. They have also begun to share a common underlying base of components, machinery, and materials technologies. There are several significant implications. First, by spreading the huge development costs across many more units, high volume markets can support the development of advanced technologies previously initiated only by military spending. Second, price-sensitive consumer applications demand that the unit cost of the underlying technology components be very low. For example, auto producers will pay an order of magnitude less for semiconductor component technologies than would contractors applying the same or similar products to military systems. Low consumer product costs cannot be achieved by reduced functionality or reduced reliability, since, for example, a real-time processor for engine or brake control on an automobile is a very sophisticated element incorporated in systems that must not fail in operation. The necessary low costs can be achieved only by the scale, scope, and learning economies of the revolutionary production approaches detailed in the previous section. The end result is that new, militarily-relevant generations of cheaper but sophisticated and reliable technologies emerge from high-volume commercial markets.

Moreover, the new production model's emphasis on speed of product development and rapid cycling of technology introduction has additional, critical military consequences. Using the strategies and production capabilities of the new manufacturing, Honda and Toyota can now take an automobile from design to showroom in less than three and one-half years. This is twice as fast as traditional mass production even though, with the incorporation of electronics and other new technologies, automobiles pose highly complex systems development problems akin to military product development (albeit with different performance parameters). Imagine the implications for military system development, plagued as it is with cycle times that incorporate technologies often two generations old, technologies that are advanced as design begins but old by the time production starts.[86]

It is a plausible hypothesis that civilian developers who have mastered the new manufacturing can move complex systems from design to battlefield faster than traditional military suppliers. They are better organized to do so. The very concept of the fastest route to the most advanced but reliable military systems in the field may have to change. The quickest route may no longer be to jump to the extreme limits of the technically feasible at the moment a system is conceived. Rather, the most effective route may well be the iterative innovation that Japanese firms have mastered.

Product development done through an endless series of small innovations may not be heroic. It can nonetheless outpace product development that attempts to jump dramatically from one frontier to the next. As IBM's former chief scientist, Ralph Gomory, put it:

> The process of repeated incremental improvement . . . an existing (not new) product gets better and develops new features year after year. Though that may sound dull, the cumulative effect of these incremental changes can be profound. . . . If one company has a three-year cycle and one has a two-year cycle, the company with the two-year cycle will have its process and design into production and in the marketplace one year before the other. The company with the shorter cycle will appear to have newer products with newer technologies. But, in fact, both companies will be working from the same storehouse of technology. *It is the speed of the development and manufacturing cycle that appears as technical innovation or leadership. And it takes only a few turns of that cycle to build a commanding product lead.* [Emphasis added.][87]

Of course, given its incremental character, the limits to the cyclical development approach will be reached whenever the development of wholly new technologies is needed. Nonetheless, the new approach shows every prospect of producing a wide range of established military systems with equal or superior technology and capabilities, but faster, far more cost-effectively, and with greater reliability in the field.

Will America Adapt?

The American military technology system is not well positioned to accommodate the new alternative—neither to integrate the new high-volume technologies into military systems in a timely manner nor to support the commercial development on which military technology now heavily depends.[88] Neither the financing nor the organization of the American R&D effort adequately comprehends the emerging reality of high-volume commercial technology development. Although the level of American R&D remains high, expenditure is dominated by military needs. Conversely, by international standards, the civilian effort is low, and the part financed privately is very low.

Similarly, traditional concerns about diversion and overpricing of scarce resources (and about controls on the commercial diffusion of dual-use technology) gain new significance when confronted by directly competing commercial technology development efforts abroad. National scientific and engineering resources are limited at any given moment. Government funding of military applications can divert essential personnel from civilian efforts and make them more expensive by bidding up salaries. This can make commercial efforts less cost-competitive and can even push domestic firms to move development efforts offshore. Similarly, dual-use restrictions and export controls discourage firms from leveraging commercial markets for military technology development. In a world of government-sponsored commercial R&D efforts, commercial products all too often are developed outside the United States from the same generic technology as that which underlies American military systems. The spin-off approach may thus delay rather than facilitate both commercial position and mastery and improvement of militarily relevant technologies.

Perhaps more important for security, the old system has created a domestic military-industrial enclave that is profoundly unlike the commercial world and organizationally unprepared for the emerging competition. Project bidding and accounting procedures involve selection criteria that amount to highly politicized speculation about future cost, performance, and procurement, and inherently limit incentives to develop the most cost-effective technologies. The consequent mechanisms installed to control abuse compel highly bureaucratic management approaches. Indeed, firms dependent on the military for research and production contracts adapt their organizational structures to market to the Pentagon.[89] This leaves them with business strategies and organizational structures ill-suited to the commercial world.

Civilian and military initiatives represent two different ways of developing advanced technology. Technical sophistication, high reliability, low costs, faster development cycles, and flexibility characterize the emerging commercial-based trajectory, best represented by the Japanese model. It is a trajectory rooted in a different community and market context than the military-spending trajectory that still dominates in the United States—a "spin-off" trajectory that, as Jay Stowsky argues in Chapter 4, no longer works. Besides the obvious advantages of cost and efficiency, this commercial-based trajectory may be better suited than traditional military spending to respond to the unpredictable regional conflicts that are likely to characterize the next century. This trajectory would also produce a very different technological foundation for the U.S. security position.

For the United States, the shift from spin-off to spin-on and the potential conflict between commercial and military trajectories pose severe policy questions. Is the American approach to military development obsolete for its own purposes? Is it counterproductive for the long-run development of the national industrial base on which militarily relevant technology development rests? Military spending and military technology development are not going to rescue the civilian economy from its competitiveness problems. Nor can they ensure sufficient national technological development even for security purposes. As the next section argues, these are disheartening conclusions for an America slowly ceding its influence to regional autonomy.

IV. Autonomy with Interconnection: The Regional Economic Structure of Security

The preceding sections have argued that U.S. capacities have declined and new capacities have arisen abroad. This section contends that, as a consequence of those developments, new patterns of economic dependence and autonomy are emerging which amount to a fundamentally new industrial foundation for security: The international economy is becoming a multipolar system organized around three distinct regional groupings. This alters not only the American security problem, but the very structure of international politics as profoundly as the changes in Eastern Europe or developments in the Gulf.

Enduring national power rests on the capacity to respond to external challenge by marshalling economic, technological, and military capacities to support national goals. When the global distribution of technological and industrial capacities changes, national patterns of external dependence and autonomy also shift. A basic change in a nation's capabilities to provide for itself shifts its rank in the international system. Sharply diverging rates of industrial growth, or technology developments that displace established military systems, can quickly change relative national positions. Consequently, national economic capacity itself must be understood as a dynamic concept, capable of significant shifts over time.

A nation's economic capacity is in great part a function of its internal political economy. That domestic economic capacity to sustain industrial and technological position underpins the capacity to offset external threat by internal development or reinforcement of the nation's international capability.[90] Measuring a nation's dynamic economic capacity for industrial adjustment, measuring thereby its capacity to maintain its ability to engage in internal balancing of external threat, requires a look inside the nation-state, at the root of its dynamic capacity, and at how that dynamic capacity may be evolving. Consequently, in the present discussions, labels such as "superpower" ought to be avoided. Such terms embody in advance a definition of who is capable of internal balancing, of acting politically to extend economic and technological resources to respond to external challenge. They can mislead us about our choices for the future.

The structure of the international system—the distribution of national capacities—has changed. The purposes to which the new capacities will be put are yet to be defined for Europe or Japan, and may be defined anew for the United States. The alliances formed to pursue as-yet-undefined threats are not evident. But if our argument is right, the international system that emerges in the next decade will be very different from the one constructed by American hegemony, and perhaps much less congenial to U.S. interests.

A Multicentered Global Economy

A more global international economy is visible in trade, direct investment, and finance. Products, companies, and investments from each of the major industrial regions can be found in almost every market on earth. International financial markets of enormous scale and significance have emerged over the past twenty years; yet, global wholesale banking rests firmly on national foundations, and retail financial models remain national. There may be a more global international economy, but that does not end the importance of place—community, district, nation, or region. Economic strategies and responses to new competition are generated within particular places, rather than by world corporations that stand outside a home base. Multinational corporations may someday be able to act without national constraint, but not yet. Firm strategies and tactics are formed within particular institutional arrangements and supply bases that at once constrain and direct their choices.[91]

There are three distinct, though interconnected regional economies, each with its own economic and technology base: Asia, North America, and Europe.[92] The United States/Canada and Western Europe each represents about 25 percent of global GDP. In 1987, Japan accounted for 12.4 percent, and Japan plus the East Asian newly industrialized countries (NICs), 15.8 percent, of global GDP.[93] The latter region will continue to expand in relative size because growth rates in Japan and Asia are substantially higher than in the United States or Europe. Direct foreign investment by Japan over the last decade in Asia has constructed a Japan-centered industrial economy and pushed the United States out of its position of pre-eminence.[94] Europe's move toward greater integration with the 1992 plan, and its financial and political concomitants (e.g., the EMS, perhaps a European Central Bank), are creating an equally autonomous region. But weighing the evidence to select between a proposition of globalization and a hypothesis of regional separability, or trying to measure how far we have come along the road to a global economy, misses the point. We must interpret the pattern in the fabric in order to understand how international markets are interwoven into regional economies.

Consider foreign direct investment (FDI) and the much talked-about role of the border-leaping transnational corporation (or multinational corporation—MNC). In the view of some, the emergence of the transnational corporation is creating an integrated global economy. Indeed, foreign direct investment grew much faster than world trade between 1983 and 1989, expanding at a rate of almost 30 percent compared to under 10 percent for world exports.[95] Roughly 80 percent of the flows during this period were within the three major regions, suggesting a further integration of the advanced countries. But if we look closer, the regional pattern reemerges and hints of American weakness appear. First, as Sylvia Ostrey notes, "a significant aspect of the 1980s FDI wave is what appears to be the emergence of regional strategies by the triad's MNCs, leading to the likely formation of investment blocs and thereby also hastening intraregional trade integration. The clustering pattern which is emerging among the countries shows each region dominated by investment from a single triad member: the Americas by the United States; Asia by Japan; and Eastern Europe as well as selected African countries by the E.C."[96] That is, the transnational corporate investment flows are themselves shaping three global regions. Second, even as three regions are being created, Japan (principally) and Europe (secondarily) have spent the 1980s entrenching themselves in the American market. The United States has become the most prominent recipient country for FDI, in part the mirror of its huge trade deficits, receiving almost half of the annual flows throughout the decade. Japan has become the principal source country, also a mirror of its large and entrenched trade surplus. So, much of the flow of Foreign Direct Investment reflects the changing position of the United States from source to recipient, its transformation from a country whose companies used their technology and organizational advantages to implant themselves in host markets into a host country itself.[97] Thus the dominant American position in investment, as in technology and trade, has receded.

The economic interconnections between the regions should not be exaggerated. Nations have long been vulnerable (that is, unable to reverse their sensitivity) to developments outside their borders and to international market exchanges outside their control.[98] Critical vulnerabilities, those that threaten the stability of the political regime or the economy, are not new either. Rather, the issue is whether sensitivity (and especially critical vulnerability) to developments outside immediate control can be reduced or countered. Although economic interconnections have grown, so has the capacity of national governments and industries to respond. National capacities to prevent interdependencies from threatening the regime or economic stability have grown even faster than the interconnections themselves. For example, compare the advanced countries' capacities to respond to external shock and stock market disruption in the 1970s and 1980s with the economic and political dislocations of the 1930s.

However, national capacities to respond to and shape ties to the international system, and to adapt domestically in response to international challenges, vary dramatically. They vary with both political and economic constraints. National capacities to establish position in the global system are a function of (1) the relative size and power of the national economy and (2) the political and administrative capacities of the national government. Therefore, the critical issue is not the extent of interconnections but their structure (which countries hold the strongest position), and how nations respond to their character. (Some interconnections pose greater problems of vulnerability than others.)

The structure and character of interconnection among the three regions will be fixed by policy choices as much as by market dynamics. National and regional differences will shape the character of international trade and investment flows. For example, Japanese firms can obtain American technology and know-how by acquiring U.S. firms, but such acquisitions are virtually impossible in Japan. Many European countries are attempting to shape the impact of foreign direct investment with a variety of policies including local content requirements. The United States is not. Europe and Japan are both seeking and increasingly establishing independent technological bases. They are attempting to ensure the foundations of national autonomy through domestic action. The implications for the system as a whole are considered in Chapter 6.

The conviction is widespread in Japan that it will be the dominant technological power by the end of the century, if not before.[99] European governments, the Community, and major European companies are increasingly investing the resources required to overcome existing weaknesses and play to technological strength. There is a growing belief, almost a conviction, that Europeans can reestablish themselves as leading players on the world stage. Meanwhile, the U.S. government assumes that market development will ensure its future position in technology and industry.

Each region has the capacity for internal balancing and, the existing resources to expand its national or regional capabilities as a response to external threat. Japan with political capacity has created economic resources; Europeans with extensive underlying economic resources are creating the political capacity to exploit them. The suspect case is the United States. Our concern is that America is substituting dependence for dominance, while thinking it is establishing an interdependent world of managed multilateralism.

The Asian Economic Region

Consider first the Japan-centered Asian trade and investment region. By almost any significant measure Japan, rather than the United States, is now the dominant economic player in Asia. Japan is the region's technology leader, its primary supplier of capital goods, its dominant exporter, its largest annual foreign direct investor and foreign aid supplier, and, increasingly, a vital market for imports (though the United States remains the largest single import market for Asian manufactures). Japan's own economy is decreasingly dependent on other world markets for growth. Japan's export dependency dropped from a high of 13.5% of GNP to just 9.5% in 1989, signalling the economy's reversion to its historical level of domestic demand-led growth.[100] Despite this, Japan's trade with the rest of Asia in 1989 surpassed her trade with the United States, more than doubling since 1982 to over $126 billion.[101]

Trade within Asia has grown faster than trade between Asia and other regions since 1985.[102] By 1988, intra-Pacific Basin trade had risen to almost 66 percent of the region's total trade, from about 54 percent only eight years earlier.[103] The major source of imports for each Asian economy is usually another Asian economy, most often Japan. In the late 1980s, for example, Japan supplied on average about one-quarter of the NIC's imports (versus America's 16–17 percent). Indeed, Japan supplied well over 50 percent of Korea's and Taiwan's total imports of technology products in the late 1980s, more than double the U.S. share of technology imports to either. Conversely, the NICs have increased their share of Japan's imports of manufactured products, from 14 percent to 19 percent between 1985 and 1989.[104] Over that time frame, increased intra-Asian trade has permitted the NICs to reduce their dependence on the U.S. market, with U.S.-bound exports falling significantly.[105]

Financial ties further reinforce intra-Asian trade trends. By 1990, Japanese industry was investing about twice as much in Asia as was American industry.[106] From 1984 to 1989, there was as much direct Japanese investment in Asia as in the previous thirty-three years, thus doubling the cumulative total.[107] Japanese investment in the Asian NICs grew by about 50 percent per year, and by about 100 percent per year in the countries of the Association of Southeast Asian Nations (ASEAN). Perhaps even more indicative, in several emerging Asian economies, cumulative NIC direct investment in the second half of the 1980s surpassed the cumulative U.S. total (by as much as five times greater in Malaysia).[108] Moreover, the use in Asia of the yen as a reserve currency is expanding sharply.

The result of such trade and investment trends is a network of component and production companies that make Asia an enormously attractive production location. That regional production network appears to be a hierarchy dominated by Japan. Japanese technology lies at the heart of an increasingly complementary relationship between Japan and its major Asian trading partners. Japanese companies supply technology-intensive components, subsystems, parts, materials, and capital equipment to their affiliates, subcontractors, and independent producers in other Asian countries for assembly into products that are sold via export in third-country markets (primarily in the United States and other Asian countries).[109] Conversely, nonaffiliated labor-intensive manufactures and affiliated low-tech parts and components flow back into Japan from other Asian producers. Summarizing these trends, MITI noted in 1987 the "growing tendency for Japanese industry, especially the electrical machinery industry, to view the Pacific region as a single market from which to pursue a global corporate strategy."[110]

As noted above, Japanese investment seems to be pursuing that strategy with a vengeance. In auto-making and electronics, there appear to be two key elements to the strategy. One is to spread subsystem assembly throughout Asia, while persuading each government to treat subsystems originating in other Asian countries as being of "domestic origin."[111] The second element is to keep tight control over the underlying component, machinery, and materials technologies by regulating their availability to independent Asian producers and keep advanced production at home. The two elements together would tend to deter too rapid a catch-up by independent producers to the competitive level of leading Japanese producers, while simultaneously developing Asia as a production base for Japanese exports to the United States and Europe to avoid bilateral trade frictions.

In sum, advanced products and most of the underlying technologies are thus dominated by Japan, with labor-intensive and standard technology production in the periphery of the region and often under the control of Japanese industry. As a result, there is resistance to these patterns by other Asian countries. In a sense there is a competition of corporate and national development strategies. The Koreans seek to break their technological dependence with national technology programs implemented by the large *chaebol* firms. The Taiwanese, Thais, Malaysians, among many others, marshall policy to their local circumstances in an attempt to reshape the existing regional division of labor. To some extent, all of the region's economies seek to emulate some of the developmental policies and business strategies responsible for Japan's success. But the developmental competition is likely to reinforce Asian autonomy even if it relaxes Japan's control over the division of labor.

For the foreseeable future, though, the character of Japanese development and policy is crucial to an understanding of the region's potential for autonomy. Modern Japanese history is in a sense the story of the self-conscious pursuit of economic development as a means to respond to external constraint. The Meiji restoration, marking the beginning of modern Japan, was a response to the threat of foreign intervention. The creation of the modern state established the political will and instrument to generate an economic transformation; the Japanese bureaucracy then acted strategically to create a market system, the conditions for rapid growth, and industrial/technological development. Since World War II, strategic economic development has provided a foreign policy tool for nations who could not achieve influence in the international system by force or threat of force. As Section II described, Japanese industry and government policies of protection and promotion acted together to restructure the domestic economy and to create competitive advantage in global markets and comparative advantage in ever higher valued-added and technologically advanced industries.[112] In essence, Japan shaped the character of its links to the international economy as a means of changing its place within the international system.

The basic elements of an autonomous development strategy are still in place. As Japanese firms have become dominant in some sectors in world markets, the Japanese economy has become more open. However, in the advanced-technology sectors, old patterns have continued, especially domestic closure combined with intense internal competition to develop products and technologies originating in Japan or borrowed from abroad. Relative autonomy is readily apparent in trade, investment, and technology.

In trade, for example, Japan still tends not to import in sectors in which it exports and, despite progress, its overall level of manufactures imports is still quite low.[113] Although manufactures have doubled to account for about 50 percent of Japan's imports, that is still far below the level of the United States and Germany, each with 75–80 percent. Moreover, the recent upsurge in imports is at least as much a story of the regional adjustment of Japanese industry to the yen shock as of the opening of the Japanese economy. Quantitative studies of Japanese imports suggest that in technology-intensive sectors, imports are tied to Japanese firms, a finding backed up by MITI surveys indicating that perhaps half of manufactured imports reflect intrafirm transfers between Japanese companies and their affiliates in foreign countries.[114] Comparing equipment purchases by subsidiaries of Japanese, European, and American firms in Australia is likewise revealing.[115] European and American firms buy equipment widely on global markets; Japanese firms buy almost exclusively from Japanese suppliers, returning to Japan for equipment.

Nor is Japan fully open to direct foreign investment. Though Japan is an increasingly prolific foreign investor, it has not permitted comparable foreign ownership of its domestic economy. Restrictions on takeovers, while serving the important domestic purpose of maintaining social peace and order, are still enormous barriers to foreign investment. Though direct investment into Japan has increased substantially over the past decade, by the late 1980s foreign direct investment in manufacturing accounted for less than 1 percent of Japanese manufacturing sales, employment, and assets. The comparable figures for the United States and Germany were 7–10 percent and 13–18 percent, respectively.[116] Finally, as Section II argued, although technology and advanced know-how flow easily from the rest of the world into Japan, they do not yet flow as easily out of Japan, except as embodied in Japanese product exports.

The asymmetry of access to technology, markets and investment opportunities is substantial whatever the mix of causes—policy, market structure, business practice, or consumer preference. Asymmetrical access maintains a strategic advantage that guarantees Japan far more autonomy in development and a sound capacity to respond to external constraints. Foreign firms enter licensing arrangements they would not consider either in the American or European market. Where once the government forced technology licensing (and foreigners accepted it because they perceived Japan as weak), now financial muscle and market strength ensure a flow of foreign technology into Japan. The insulated domestic market permits firms to compete intensely among themselves, honing product and processes, and then pour exports onto foreign markets. Other countries are then forced to absorb the excess capacity that Japan's market-share strategies generate. Together, asymmetrical access and overbuilding of capacity result (as, for example, in semiconductor technology) in Japanese developments' precluding or slowing the commercial development of the technology by foreign producers. This strategic advantage can be demonstrated both in particular sectors and across industries.

Japan's relative autonomy and capacities are further enhanced by the emerging economic architecture of the broader Asian region. As argued earlier, Japan is at the core of a region of vibrant and rapidly expanding countries. Networks of excellent production capabilities exist throughout the region, attracting producers from outside to relocate in Asia—not, as before, because the shop-floor workforce is cheaper, but because the workforce is better trained and the engineers are cheap. Once production is transplanted, new product and technology development tends to follow.[117]

In sum, for the last half-century Japan has acted self-consciously to build its industrial and technological foundation. It continues to act to balance external weakness with internal action. Now a production core essentially independent of American technology and know-how (though tied for the moment to American markets) is emerging in Asia. As Asian incomes rise, a growing Asian market may further disconnect Asian growth from its tie to the U.S. market. Emerging understanding and emulation of Japan's success guarantee that Japanese innovations in policy and corporate strategy—and eventually in manufacturing as well—will spread throughout the region. The prospect for Japan and the rest of Asia is for increasing autonomy conjoined with real capacities to handle constraints arising from outside the region.

The European Region

The indicators of growing Asian regionalism find a counterpart in Europe. In Europe, though, there is an overt political as well as economic dimension to the story. We examine Europe's evaluation in greater detail in Chapter 3. Economic and political challenges have pushed the national powers of Europe to consolidate their markets and their influence. The movement to create a single European market is driven not only by the emergence of Asia, but by the perceived real decline of the United States as a source of technology, production know-how, and hegemonic influence. European elites are rethinking their roles and interests in the world, reconsidering their relations with the United States and within the European Community. That movement was then accelerated by the disintegration of the Soviet Empire.

For the last two generations, Europe's economic position has rested on a set of implicit bargains with the United States in technology, finance, and trade built inside of the explicit security bargains.[118] In technology, Europe could not lead but it could still acquire it relatively easily from the United States. Though Europe trailed in development, it excelled at applying advanced technology. Its position of privileged second may have grated a bit and did induce efforts to build capable national champions, but it was tolerable and did not provoke united European action. In finance, the dollar anchored the international financial system. That provided privileges to the United States, but stability for others and, at least until 1971, the right to devalue against the dollar to maintain trade equilibrium. Thus, if Europe could not structure financial rules to its liking, it could at least adjust to American positions. In trade, the United States maintained an open market and encouraged the creation of the Community.

In technology, finance, and trade, in sum, if Europe was not first, it was second, and individual bargains by and between European governments and companies sufficed to generate economic growth and significant geopolitical influence. Over the past fifteen years, however, that situation changed dramatically. Japan's rise and America's decline meant that Europe's position would become even more constrained: Suddenly Europe has to stomach the prospect of being third. Crucial technologies now often appear to be available only from Japan; Tokyo and Bonn as much as Washington shape financial evolution; and in trade, American legislation and bilateral arrangements threaten to disadvantage European industry, while the Japanese market remains relatively impermeable.

Set aside arguments about culture or history. America and Europe share a security structure, but Europe and Japan do not. For the Europeans, to be even modestly dependent on Japan in finance, trade, or technology—is unacceptable without the integrated defense and economic ties that link the Atlantic partners. Asymmetrical access in technology, investment, and trade without integrated security ties makes exchanging a hegemonic America for a hegemonic Japan wholly unattractive.

With the retreat of Soviet power from Eastern Europe and the reunification of Germany, an abrupt reorganization of Europe has confronted the ongoing EEC process. Although these political developments initially risked splintering an emerging Europe back into squabbling national powers, they now appear to have generated an increased commitment to the European project. The clear evidence of that increased commitment can be found in the agreement to pursue European monetary union, which almost certainly will reinforce German economic leadership, increased political union, and the effort of the European Free Trade Association (EFTA) members to negotiate increased accommodation. At least some of the reasons seem evident. A reunited and increasingly powerful Germany can be safely anchored only in a strengthened European community. NATO always provided two containments, an overt containment of the Soviet Union and an implicit containment of Germany. The EEC was founded in part to serve as an anchor for Germany in the West. Now as NATO recedes in political significance, the Community's economic and political bargains may be recast to ensure that a reunited and sovereign Germany remains an integral part of Europe. The limits on these deals struck by governments and governing elites may be set by the willingness of the several nations to ratify the bargains or to move to implement them.

On the economic front, Europe already exists as a relatively self-contained unit. Rather than the image of a set of small and medium-sized countries increasingly open to the global economy, Europe should be seen as nations (including the EFTA countries) that have successfully moved from interlinked national economies to an integrated regional economy. Trade within the EEC has grown faster than the trade between the Community and the rest of the world since the establishment of the European Community in 1958. From 1967 to 1987, the ratio of EEC–EEC exports to EEC–non-EEC exports rose from .79 to 1.15.[119] Moreover, intra-EEC trade has been a dominant proportion of each member nation's trade. Discounting intra-European trade, Europe's percentage of world exports and imports drops dramatically: exports from 44.6 percent to 13.8 percent and imports from 42.6 percent to 11 percent.[120] Add the EFTA–EEC trade and the picture becomes even clearer. In 1967, intra-European trade accounted for 50–60 percent of Europe's total trade; by 1987, the intra-European trade accounted for 60–75 percent.[121]

These trends are likely to continue with the creation of the Single Market and the adherence of the EFTA countries to it whether they formally join or not. As in Asia, financial ties now also reinforce regional trade ties. The European currencies are increasingly bound to each other through the formal mechanism of the EMS and the predominance of the D-mark. The EMS mechanism encourages regional integrity by providing greater stability for each national currency. Progress is also being made toward formal coordination of fiscal and monetary policy, which could eventually culminate in a European Central Bank. Even the British, initially so recalcitrant under Margaret Thatcher, are now committing to increased monetary integration.

Europe's regional capacities and fundamental strengths have often been underestimated. They rest in an educated and highly skilled workforce, a sound foundation in science, and the enormous wealth built up through a long and successful industrialization. Europe's overall industrial position is strong despite the years of supposed sclerosis. Industrial strength is reflected in traditional and scale manufacturing—from textiles to chemicals—and in manufacturing equipment and materials. In these industries, European firms have been very effective, often at the forefront, in applying advanced technology to hold market position—much more successful than their U.S. counterparts. And new strengths have been added to this older foundation. Those include the continuing application of advanced technology to traditional industries, a capacity at systems development and integration, and the use of political will to retain final product markets in the face of production or product advantage.

The most obvious weaknesses of the postwar years are now being confronted (e.g., the failure to be competitive in the range of the advanced electronic products from semiconductors to computers). Some of the programs, such as those in telecommunications, are likely to succeed; others face real difficulty. But through a variety of mechanisms ranging from subsidies to management of direct investment, the Europeans are attempting to maintain, and in some cases rebuild, essential capacities. This is particularly true in electronics, where a combination of changes in trade rules (e.g., rules of origin shift from assembly to fabrication in the chip industry) and novel enforcement (e.g., tying of dumping to local content) amounts to an explicit policy to force foreign direct investment to rebuild the local electronics supply base. The subsidies flowing into electronics and information technology are enormous (e.g., $3–$4 billion ECU just for semiconductors). The first round of community programs that focused on direct support for procedures are being reconsidered. The latest conceptions and language emphasize market forces and leading-edge users as a means of promoting advantage. This rethinking probably presages a shift in the emphasis of technology development programs and perhaps of trade policy to favor the needs of users over producers. But while the tactics and perhaps strategy may change, the objective remains firm.[122]

Europe is by no means a single political actor. It will remain a set of national, political communities, and as a region, a bargain among governments. Nonetheless, in a growing number of domains, including trade and, increasingly, finance and technology, European governments are able to act jointly to create regional capabilities. In a world of autonomous regions, Europe would have significant advantages—not least collective wealth, size, education, and political will. Will distinct European security interests emerge? Will Europe as a community pursue a distinct international strategy? There is a growing conviction that Europe can reestablish itself as a leading global player, building capacity for independent action while minimizing perceived vulnerability and dependence on the choices of those in other regions. Like Japan and the rest of Asia, Europe appears to have both an industrial/technological base capable of providing for itself and the political will to maintain that capacity and respond to external constraints. How should the United States react to these developments?

V. Converting Economic Power into Political Influence: Toward the Next Security System

A new distribution of economic resources, of industrial and technological capacities, alters, almost by definition, the constraints and choices for the major nations. Section I explored U.S. decline and the development of nascent dependencies in important areas of trade, finance, and technology. Section II situated

U.S. decline as the counterpart of new and powerful industrial capabilities emerging abroad, especially in Asia. Section III suggested that those capabilities were shaping a new technology development trajectory with real implications for security—a new technological foundation for security. Section IV explored the emerging regional distribution of industrial and political capabilities, suggesting how an alternative economic foundation for a new security system could be emerging. Do the developments explored in the first four sections point to a manageable multilateral security system with continued U.S. leadership, or to something entirely different and less congenial to U.S. interests?

New resource distributions do not define purposes and interests, let alone alliances and rivalries. Japan's current conception of comprehensive security emphasizes autonomous finance and trade capabilities but includes continued dependence on an American military umbrella. Similarly, Europe is not a unified protagonist in foreign and security affairs (although in finance and trade it is rapidly becoming one). Nonetheless, the retreat of Soviet power from Europe, the Persian Gulf aftermath, and political changes in Asia all leave Europe, Japan, and the United States with different concerns. For example, if major emigration is provoked by upheaval in the Soviet Union and the Gulf, Europe—not the United States or Japan—will be the primary destination. As the security problems shift, become differentiated, and take new forms, each actor will redefine its visions of security and conceptions of interests. Predicting the evolution of the security system is impossible, but we are able to identify several issues.

Military Potential and Security Interests

At the moment, Europe and Japan are regional economic powers that have the industrial and technological capacity to put a strategic military machine in place. Although the machine does not now exist, Europe and Japan can achieve significant political leverage with the potential to move toward autonomous security positions. Their military potential and changed sense of threat may alter what Europe and Japan will pay for security.

The military potential is very substantial. As suggested earlier, Japan has the component and subsystem expertise to put into place almost any military equipment it chooses. It already builds sophisticated weaponry such as tanks and smart missiles, and is developing systems expertise in aerospace. Recall that the FSX was an American alternative to independent Japanese development of a fighter plane. Many Japanese believe that Japan could have built a better plane on its own.[123] The increasing electronics content of weapons may well provide Japan an opportunity to quickly establish an advanced weapons position by trading expertise in avionics for expertise in aeronautics.[124] Already, many contend that Japanese military electronics are more reliable, with longer intervals between service or failure.[125] Japanese industry and policymakers are quite aware that they are likely to be able to produce systems less expensively than the United States. In short, the restrained

Japanese military position comes from political choice, not industrial or technological constraint.

Europe's situation is quite different. European countries and industry can build varieties of military systems of all types—indeed, too many varieties. Combined and coordinated, Europe would be a formidable military player. Amid conflicts and doubts, there is identifiable, if tentative, evidence—in planning, procurement, and industry consolidation—of increased European commitment to common defense structures. At the moment, each nation within Europe is dependent on the United States for important technologies. In exact complement to Japan Europe is weakest in the underlying component technologies, strongest in systems expertise. Needless to say, Europe–Japan industrial alliances that are emerging could provide a formidable challenge to U.S. leadership in military systems—even if such collaboration arose only as a consequence of joint commercial projects with substantial spin-on technology. Such outright collaboration on military systems does not require a formal alliance structure: When the FSX deal was negotiated with the United States, Dassault and the French were exploring a similar venture with the Japanese. Like Japan, Europe now acts out of political choice rather than industrial constraint. Regional industrial capabilities in Japan and Europe now make possible the pursuit of autonomously defined security objectives. That these objectives continue to fall in line with traditional U.S. interests is now mostly a matter of political choice.

Just as important, as Francois Heisbourg argues, the postwar security structure was defense-oriented, clear-cut, comprehensive, and rigid.[126] In the new system, confusion and complexity will prevail, defense will lose its centrality, and the nature of threat will become ambiguous. Consequently, Europe and Japan may not need fully autonomous military capabilities to assert and defend autonomous security positions. They may achieve significant political leverage with the mere potential. The circumstances that would spark a reformulation of Japanese or European policy are diverse but could become compelling. The push could come from conflicts and civil wars, likely with the disintegration and reorganization of Central Europe. Or it might start with the Gulf War's cost and consequences. The European discussion of a common defense policy has already begun. It is a discussion in which the place of America is by no means clear.

Europe and Japan have sufficient military, industrial, and political capability to alter the security structure. A situation in which U.S. leadership and managed multilateral security continue because our allies choose not to define an alternative is radically different from one in which U.S. leadership is produced by strength. That situation persists at a time when the rapid changes in Eastern Europe and the Persian Gulf will almost certainly force a redefinition of the relations between the advanced countries, whether or not that redefinition begins within existing alliance institutions. The character of any new security system will depend on the distribution of capacities and the perception of threat, on the balance of external constraints, and on the opportunities per-

ceived in each region. From these will emerge the alliances of the next security
system.

Constraint, Influence, and Competition

Constraint and Influence: Economic and technological dominance has been the
foundation of American leadership. As we argued earlier, the ability to generate and apply economic resources to the direct exercise of power, or to shape
indirectly the international system and its norms, has always been the economic
foundation of national security. When allied nation-states are knit together into
a shared security system, power within the alliance resides with the nation that
has the ability to get the others to act on its behalf or the wherewithal to put to
its own use resources belonging to the other states. That can be accomplished
directly through overt threat and punishment (or promise and reward), or indirectly when the structure of the system channels particular outcomes that serve
the lead state's interests, or when the lead state's preferences become the
alliance's accepted norms.[127] In the postwar era, industrial and technological
resources supported U.S. military strength and underwrote the use of commercial and technical assistance to secure allied agreement with U.S. goals. But the
economic dimension was just as critical, indirectly, in its impact on the system's
structure and norms. Now, from a position of dominance, America has begun
to risk dependence in industry, finance, and critical segments of technology. It
seems to us that the situation is now reversed: American influence has not just
diminished; rather, we now risk—and the emphasis is on *risk*—dependence in
technology, economy, and finance.

There is little point in speculating on specific instances of future foreign
influence and leverage over U.S. decision making. That would require specifying the circumstances in which such leverage would be used. Rather, it is useful
to identify some categories where America once exerted influence but now risks
dependence. Consider finance: During the 1956 Suez Crisis the United States
threatened a run on the pound to constrain British, French, Israeli action. Now
Japan has the financial capacity—and begins, in public, to threaten to use that
capacity—to influence the American exchange rate and monetary conditions.
Or compare the enormous influence the Marshall Plan permitted the United
States after World War II with our relative financial inability to invest in developments in Central Europe today. To stretch the point, but perhaps not so far,
note the American use of International Monetary Fund (IMF) conditionality as
an instrument to shape domestic choices abroad—even in England. Then note
that a single European currency might well make the EC the largest bloc at the
IMF, and, perhaps, as some EEC officials gleefully note, move the IMF out of
Washington, not just out of American dominant influence.

Similarly, industrial position and financial aid have been significant levers
of American influence. Moreover, the success of the American industrial model
as one to be imitated reinforced U.S. preeminence. Now Japan dominates the
Asian economic region. European market position in electronics and autos has

suddenly made Japan and Japanese companies players in Europe. The Japanese "lean" production model calls for emulation; the American system suddenly is depicted as a rigid past.

Finally, as argued here at length, there is technology. America has often shaped decisions of other nations by denying or threatening to deny them access to technology. Why should we expect our experience of technological dependence to be different? And dependent we are in critical electronics and production equipment technologies.

With a reduced military threat, the security problem takes new form. The question of how to achieve security must be redefined. Traditionally, control and use of resources required armed force. Now resources can be controlled through markets. The security issues do not disappear, but they are submerged and hidden by market and social relations. Market structure and functioning must become a matter of direct security concern. The forms it may take will be diverse.

Competition: In a world of competing regional economies, a series of economic relations may be defined and perceived as threatening the ability to preserve the community within a particular territory from outside intervention and control. In that case we would expect nations, or regional governments, to act in economic arenas in ways reminiscent of their security strategies. The stakes may well be the relative course or trajectory of economic development. Whether one country can act to shape another country's trajectory of development is critical to determining whether foreign actions will be seen as security threats or simply irritants on the trade front. We return to this problem in Chapter 6.

There is an emerging intellectual basis for interpreting the particular trade, technology, and investment frictions among competing regions or nations as security threats. A multipolar security system built around a world of mercantilistic competition can be conceived, and justified from increasingly established reasoning. Two analytic frames are necessary—the technology trajectory arguments developed earlier and so-called new-trade theory.

The new-trade theory argues that in oligopolistic industries governments can reshape the structure of global competition and global industry to the benefit of national welfare.[128] National gains can occur under two conditions. First, imperfectly competitive industries (e.g., characterized by oligopoly) tend to earn higher returns (excess profits or rents) than those available in other sectors of the economy (where competition bids away excessive returns). Policies that help to win larger market shares for domestic producers in these industries will increase national welfare at the expense of other countries by capturing a larger share of the global profit pie. Second, and more important, certain industries generate external economies (i.e., social gains far in excess of capturable private returns). Government policies to promote or protect these industries can improve welfare by fostering and capturing these spillages. High-technology industries are likely to fall into this class because of the broad knowledge generated by their R&D, and because of the price/performance improvements they create in the industries that apply them. Since most major industrial sectors consist of a limited number of large powerful firms (oligopoly) and since high technology increasingly occupies center stage in trade disputes, the new-trade theory really addresses the core of industrial competition among the advanced countries.

If we marry the implications of the new trade theory to those of the technology trajectory arguments developed earlier, the result is explosive. Then, the outcome of strategic trade conflicts is not simply a matter of one-time gains or losses that result when one government's policies assist its firms to gain share in global markets to the disadvantage of its trading partners. National position for particular firms in their markets is not the only issue; nor is the current position of one nation in the international economy its final reward. At stake are future gains and losses in terms of each nation's dynamic potential for long-term growth, increased standards of living, and technological preeminence. Trade and domestic technology strategies quickly become the stakes in international conflict.

The more these theoretical arguments are accepted, the greater are the perceived stakes in individual trade disputes that influence technology development. This is not simply an intellectual puzzle. The new theories and their marriage simply express or give foundation to an intuition that is, in any case, driving policies abroad. Our concern is that trade debates will be seen as direct security issues. The risk is that the language of security conflicts will be recycled for trade debates.

There are few useful guides for exploring these new security issues. The real difficulty is not which problems existing theories treat or how, but what they do not effectively address. Theories of interdependence do not centrally confront the problem of how market processes, and government's manipulation of market processes, shift real power from one nation to another. The studies in international political economy that consider the interplay of state and development focus on the emerging countries, not the advanced countries.[129] Their insights about the ties between domestic development and the international structure are only indirectly integrated into core debates about the dynamics of the advanced industrial democracies. Indeed, the International Political-Economy debate about the advanced countries has presented a problem to comparative politics: how to use national-level variables to account for variations in national economic responses to common international economic problems.

Economics as a Security Problem: The issues raised in this chapter suggest that a central problem for international political economy must be how international regimes or the structure of interdependence is manipulated by nations to create advantage. What matters is not simply order, but order on what terms. Nor is it simply a matter of which nations are makers of or takers from the system. The question is: how do dominant powers use their influence in shaping the rules of play to gain advantages in the trade and financial system. In a world of nation-states, it matters what is produced, by whom, and where. The national production profile represents both a set of economic possibilities and security conditions. A nation rich from oil, timber, and agricultural exports has vastly different potential for economic growth and vastly different security constraints than those of a nation wealthy from production of computers, advanced components, and new materials.

Industrial structure and the dynamics of technological evolution have become necessary intellectual foundations for the student of international politics. Success in trade is not simply an alternative to a security strategy.[130] Trading nations have lived in very particular balances of military power. More importantly, a trading strategy can serve as a means of creating the wealth to provide security directly, as Japan's emerging military potential suggests.[131] The international political economy debates of significance must be about the central stories: how wealth and capabilities are created and redistributed; how regimes were arranged not simply to provide order but to extract resources; and the place of political economy in military and security policy.

The Next Security Era

Although a new security era is upon us, the current security debate is still rooted in the past.[132] It has been an argument about the level and form of American contribution to a western security system with America at the center and its allies ceding the definition of crisis and response because they are dependent on U.S. action for their own security. The new reality confronts us in pieces—in fragments and isolated controversies, not yet as a whole. The reality is that our major allies have the range of capabilities required to act on their own in the international system, to behave as great powers. The reality is that the possibility of American dependence on our allies in a range of significant policy arenas is growing. Whether they use their capabilities to pursue their foreign policy preferences is increasingly a matter of their political choice.

The U.S. economy is no longer so disproportionately large or so distinctively structured around advanced production and technology as to create a fundamental foreign policy advantage. The domains in which the United States used to exert influence to extract security compliance—trade, technology, finance—can no longer provide us leverage. Those former domains of action have become binding constraints in their own right. Industrial innovation is no longer the preserve of the United States. The areas of significant industrial weakness are extensive and growing. Financial power rests in institutions outside the United States, though for the moment the system is still organized around the dollar and American-dominated international institutions.

Nor, as we have emphasized throughout, will U.S. security preeminence be maintained by current military systems advantages or a new focus on rebuilding the defense industrial base. The dominant U.S. model of military technology developed through military spending faces a less costly, more reliable, more responsive commercial alternative. The architecture of the domestic supply base is consequently shifting from autonomy to dependence on regions whose political interests may diverge from those of the United States. If present developments go unchecked, thorough-going dependence on foreign sources of military components and subsystems will become a reality. The possibility is real that technologies only obtainable abroad will be sufficiently critical to provide leverage on American foreign policy.

The bipolar era is ending. The configuration of the international system is changing. The United States is now confronted with the problem of managing relations with two other, roughly equal regions, each with the capability of acting autonomously in matters of technology, industry, finance, and security. At minimum, the formation of western security policy will become more complicated. Real differences about the organization of the international economic system, as well as the risks and potentials in the remarkable events in Eastern Europe, could become the basis for serious divisions. Allies increasingly will have to be accommodated, even given primacy. At a maximum, badly diverging interests could create the basis of real conflict between the regions.

The nature of threats to the U.S. position in this multipolar world must also be reconsidered. In the world we are describing, the continued erosion of America's international economic position is a national security issue. We are past the point where America's security dominance can be exploited to impose more favorable terms of trade. Rather, we are confronted with precisely the reverse: how others can exploit terms of trade to impose dominance, how they can structure and play the international system through economic means. In that world, the only secure America is a competitively able one. The United States must regain its competitive standing in trade, technology, and finance if it wants to be in a credible position to effectively manage the changing security system.

Think what the Persian Gulf war might have been like if our adversaries had had the technologically advanced planes, night warfare capabilities, and smart weapons. Think of the military systems a combination of Japanese componentry and manufacturing skills with European systems integration know-how might produce. Then think of the emergence of autonomous regions with the political will and capability of pursuing interests that diverge from a competitively weakened United States. These flights of fancy are increasingly possible. America needs to act not from the belief that we are and can remain dominant, but from an understanding of how we can be effective in circumstances in which we no longer are.

II

THE PLAYERS

2

The Power behind "Spin-Ons": The Military Implications of Japan's Commercial Technology

STEVEN VOGEL

Three of the biggest news stories to hit Tokyo in 1987—Toshiba, SDI, and FSX—bear a common lesson. Each of these stories suggests, in its own way, that Japan's leadership in commercial technology has enormous implications for international security relations. In May 1987, it came out that the Toshiba Machine Co., a subsidiary of Toshiba Corp., had sold sophisticated milling machines to the Soviet Union, reportedly enabling the Soviets to reduce the noise level of their nuclear submarines. U.S. Congress members staged a ceremonial declaration of war on Toshiba on the Capitol lawn, smashing Japanese-made radios in front of television cameras from around the world. Many Japanese commentators viewed the whole incident as a pretext for "Japan-bashing," pointing out that the Soviets had achieved quieter submarines before they imported the Toshiba machines. More sober analysts pointed to the Toshiba incident as a demonstration of the potential power of Japan's commercial technology. This particular export may not have altered the global military balance, but another one just might.

In July 1987, the Japanese government officially announced that Japanese firms could participate in research on the Strategic Defense Initiative (SDI). The U.S. Department of Defense (DOD) awarded two contracts of $3 million each to U.S.-Japan consortia to conduct architecture studies on antiballistic missile defense in the western Pacific. Mitsubishi Heavy Industries (MHI) leads one consortium and LTV Corp. of the United States leads the other.[1] The SDI Office (SDIO) granted both consortia a second one-year extension for a "Phase III" study in April 1991. In addition, Japanese companies are participating in SDI research on Josephson–Junction microprocessors and the fabrication of artificial diamonds.[2] Japanese participation in SDI is symbolic of two important trends. First, the DOD is becoming more interested in tapping Japan's technology base for its own military programs. The Pentagon views SDI research cooperation as a way to gain access to Japan's advanced dual-use technology and, in somes cases, to gain control over this technology. Second, Japanese corporations have begun to cultivate the DOD as an important customer for the future. Electronics companies such as Mitsubishi Electric and NEC were particularly eager to pursue involvement in SDI.

In October 1987, the Japan Defense Agency (JDA) decided that General Dynamics of the United States and MHI of Japan would co-develop Japan's next fighter plane, code-named the "Fighter Support Experimental" (or FSX), based on the American F-16. In doing so, the JDA disappointed both the Americans, who wanted the JDA to buy an American plane off the shelf, and the domestic defense industry, which was pushing for indigenous development. In November 1988, the two governments signed a memorandum of understanding (MOU) whereby U.S. manufacturers would be guaranteed 35–45 percent of development work. More importantly, the MOU stipulated that technology developed under the FSX project would flow back to the United States. The FSX issue returned to the headlines in 1989 as a group of congressmen tried to pressure President George Bush to renegotiate the MOU. These congressmen argued that Japan should buy an American fighter, given the enormous trade gap between the two countries. They also expressed concern that Japan might be able to use its advantage in important high-technology sectors to launch a full-fledged drive into the global aircraft production business. As it now stands, the FSX co-development project provides a crucial precedent in at least two respects. First, it represents the first large-scale attempt at United States–Japan joint development in the military sphere. Second, it gives Japan a golden opportunity to apply its commercial technology expertise in a military project and to assess its own system integration capabilities.[3]

Japan plays a bigger role in international security today because it controls a resource that is critical to military strength: high technology. Many Japanese themselves accept that their country is now a major power in global affairs, yet they often deny that this power has a military dimension. A report commissioned by Japan's Economic Planning Agency suggests that Japan in 1985 already had surpassed the Soviet Union and the major European powers to become the world's No. 2 power in terms of its "ability to contribute to the international community."[4] Japanese leaders insist, however, that although Japan may be an international economic power, it will never be a military power. Japan may demand more influence within the Asian Development Bank or the International Monetary Fund, but it will never claim to have much impact on the global military balance.

This chapter suggests, on the contrary, that Japan is already a major military "actor" in the world, if not a military power in its own right.[5] In particular, Japan has gained leverage over its closest ally, the United States, by virtue of its strength in high technology. If Japanese firms take the "Toshiba" route, selling technology to America's rivals, they could undermine U.S. efforts to maintain a technological edge. If they pursue the "SDI" route and transfer their dual-use technology to the United States, these firms could help the United States to maintain long-term technological superiority. This is no less important for the United States in the era beyond the Cold War, for U.S. military strategy still relies on substantial technological superiority over potential or actual adversaries, whether they be Soviets, Iraqis, or others.

The FSX project illustrates the tensions created by Japan's growing leverage over the United States. U.S. officials would like to use the project to gain access to Japanese dual-use technology, and to keep Japan dependent on U.S. military technology assistance. Japan, on the other hand, would like to use the project to try out its hottest dual-use technology and to develop system integration skills, thereby becoming less dependent on U.S. military technology assistance. If the FSX project works well, both sides may realize how much they have to gain through cooperation. If it fails, as it almost did in 1989, Japan is likely to pursue a much more independent course in military development and weapon systems production.

Japan's technological strength ultimately gives it new options for maintaining its own security. Given its technological and economic strength, Japan has the capability to become a major military power within the not-too-distant future (ten to twenty-five years). Japanese ambitions for a return to power in the world are no longer constrained by technology, but only by politics.

Japan's Commercial Technology Base

Japan's economic power, and its actual and potential military power, are ultimately rooted in the strength of its commercial technology base. Japan has built up an impressive technology base through a gradual process of adoption of foreign technology and constant innovation in methods of production.[6] Although U.S. corporations invented much of the important new technology in the postwar period, Japanese firms have been more successful in developing efficient manufacturing systems. Japanese technology expert Masanori Moritani uses RCA as a paradigm for the failure of American companies to maintain a technological lead. RCA led the way in the development of television, yet Sony perfected trinitron technology.[7] RCA was an early innovator in videotape recorders, but Sony and the Japan Victor Company refined the video recorder into a product small enough and inexpensive enough for the household consumer.[8] And RCA produced the first amorphous solar cell in 1976, but Sanyo was the first to develop it into a marketable product.[9] "In terms of commercial technology," Moritani concludes, "we don't have anything left to learn from the Americans."[10]

Japanese producers really only assaulted the heart of the U.S.-high technology advantage, however, when they began to export semiconductors. Japanese manufacturers' innovations in production technology made them particularly successful in the mass-production market for dynamic random-access memories (DRAMs). From 1978 to 1986, the Japanese share of the world semiconductor market grew from 28 to 45 percent, whereas the U.S. share declined from 54 to 43 percent.[11] By 1986, Japan had 65 percent of the world market in metal oxide semiconductor memories.[12] In the same year, a U.S. Panel on Materials Science concluded that Japan had the edge in the all-important area of processing materials for electronics devices.[13] In 1987, a Defense Science Board Task Force estimated that Japan led in both silicon and nonsilicon products.[14] Later in the year, the U.S. government and manufacturers decided to respond to the Japanese challenge by forming their own research consortium, Sematech.[15]

With their successful advance in semiconductor technology, the Japanese have essentially surpassed the Europeans and caught up with the Americans in the overall high-technology race. Evaluating an entire country's technology base is by nature an imprecise art. The results of any evaluation will vary depending on how it values the different "qualities" of technology, such as scientific novelty, technical complexity, endurance, and ease of maintenance. Nevertheless, most recent attempts to compare the overall technology base of Japan and the United States have shown rough parity between the two countries. Even the "modest" Japanese recognize this new situation of parity in the United States–Japan high technology race. The 1988 Ministry of International Trade and Industry (MITI) White Paper on Industrial Technology claims that by 1993 Japan will lead the United States in eleven of forty-one high-technology product areas including memory devices, fine ceramics, and semiconductor lasers. The United States will lead only in four areas: satellite launching vehicles, aircraft engines, databases, and magnetic resonance imaging (MRI).[16]

United States–Japan "parity" in the high-tech race, however, is not the same as United States–Japan "equivalence." In fact, Japan and the United States have very different strengths and weaknesses within what is a very tight race overall. The fundamental differences in the nature of the technology that U.S. and Japanese scientists come up with follows logically from differences in the way the two countries research and develop technology. In general, researchers in the United States focus more on basic research, whereas their Japanese counterparts concentrate more on product development. Japanese critics themselves are fond of reminding their public that Japan only has five Nobel Prize winners in science through 1990, whereas the United States has 159.[17] These same analysts, however, typically take offense at characterizations of the Japanese as "uncreative," asserting that Japanese creativity merely manifests itself in different ways. Moritani, for example, has developed a distinction between U.S. "originality" and Japanese "creativity." Although most original ideas to date have come from the West, he stresses that Japanese technicians have shown exceptional ingenuity in adapting these ideas in order to develop useful products.[18] This kind of "creativity" has been manifesting itself more and more in the marketplace. Japanese received 21.6 percent of the 90,000 patents issued by the United States in 1990.[19] Computer Horizons Inc. of New Jersey found that Japanese actually rated higher than Americans according to an index of innovation based on how often a country's patents are cited in applications for other patents. Japan achieved an index rating of 1.34, compared to 1.06 for the United States, 0.94 for the United Kingdom, 0.80 for

France, and 0.79 for West Germany.[20] Japan's "original" contributions can be expected to increase as well now that Japanese technology strategy has shifted from catching up to taking the lead.

Although Japan's relative inattention to basic research may become a greater weakness in the future, its almost obsessive attention to the subtleties in product development should continue to reap generous rewards. Japanese companies' emphasis on lowering production costs and constantly improving manufacturing technology has given many of them a significant price and quality advantage over their foreign competitors. In the future, the Japanese may be able to use their advantage in flexible manufacturing systems (FMS) to make inroads in the U.S. lead in the small-batch production of advanced electronic components and equipment. They have already achieved lower defect rates than their counterparts in mass-production sectors as diverse as automobiles and memory chips. MITI's analysis for the 1988 White Paper on Industrial Technology judges that Japanese high-technology products are more reliable than American ones in twenty-seven of forty-one areas and equally as reliable in another eight areas.[21] The implications of this reliability advantage are enormous for the future of U.S.–Japanese competition in commercial markets, but the implications in the field of military systems may be even more ominous. In the commercial market, a high defect rate can result in a severe loss of market share. On the field of battle, defects costs lives.

Japan's R&D activity suggests that Japan is now closing the gap in many areas of technological weakness and increasing its lead in areas of strength. The Japanese government and industry are committed to making the investment necessary to strengthen the country's high-technology base. Government and business leaders will not back off from competition in any significant high-technology sector despite trade friction with the United States because they see high-technology leadership as Japan's only route to long-term prosperity. Japanese leaders are painfully aware of the challenge that the Newly Industrializing Economies (NIEs) pose to Japanese heavy industries such as steel and shipbuilding and to "low" high-tech industries such as consumer electronics. They are more confident of challenging the United States in "high" high technology than they are of fending off the NIE's challenge to Japanese supremacy in "low" high technology. In any case, they would prefer to continue moving away from labor-intensive industries toward higher value-added sectors.

The Japanese government has been remarkably successful in promoting research and development, particularly given its relatively small share in overall R&D spending. In 1989, R&D expenditure for Japan totaled 10.909 trillion yen, compared to 195.960 trillion for the United States (at $U.S.=138.0 yen). The United States, however, spent 5.57 trillion yen on defense R&D, whereas Japan spent only 93 billion. The U.S. government shouldered 46.4 percent of the country's R&D burden, the Japanese government covered only 17.1 percent.[22] In part, government spending is underrepresented in spending figures because of the considerable tax incentives offered for private industry research spending. More importantly, the Japanese government has acted as an effective coordinator and facilitator of research projects without necessarily serving as the primary source of funding. The government avoids inefficient duplication of research and facilitates the diffusion of the results by aggressively promoting interfirm cooperative research, which forces private companies to share information and allows them to standardize parts more easily. In addition, joint research projects create personal networks that facilitate future cooperation.[23]

The Military Uses of Commercial Technology

Japan's present and future technological leadership is only so relevant to international security because in recent years commercial technology has advanced more rapidly than explicitly military-use technology. In the past, the requirements of the military market were usually much more strict than those of the commercial market. Products for military procurement must be resistant to shock, heat, and radiation in a way that few commercial products need be. Military technology generally was considered to be more advanced than commercial technology so that one could expect considerable commercial spin-offs from military research. Today, commercial technology is at the forefront in many areas. It is difficult to compare the levels of technology in the commercial and military sectors in any comprehensive way, but the commercial sector now leads substantially in the critical area of microelectronics. Due to the long production cycle in the defense industry, most U.S. military systems now use devices that are five to seven years out-of-date. U.S. and Japanese producers introduce a whole new generation of devices every two to three years, whereas most military systems evolve on a five-to-twenty-year cycle. The commercial market in many high-technology products has the advantage of greater size, which means greater incentives for producers and higher profits that can be recycled into more R&D. The commercial market also offers more immediate and more widespread feedback on product performance. This encourages producers to put a premium on cutting production costs and improving manufacturing processes. Finally, increased competition for reliability and endurance in commercial markets means that these products now have to be as reliable as (if not more reliable than) military-use products. A 1986 Defense Science Board report on "The Use of Commercial Components in Military Equipment" judged that commercial electronic systems such as computers, radios, and displays were just as durable in harsh environments, one to three times more advanced, two to ten times cheaper, five times faster to acquire, and more reliable than their military equivalents.[24] In the foreseeable future, commercial-to-military "spin-ons" are likely to boom while military-to-commercial "spin-offs" decline.[25]

Japanese producers are particularly likely to benefit from technological "spin-ons" because their most important area of technological strength—electronics—is becoming increasingly critical to military systems. The U.S. Electronics Industries Association estimates that the electronics content of defense

Table 2-1 Some Examples of Japanese Dual-Use Technology

Civilian Technology	Producer	Military Use
Materials		
Radar-proof ferrite paint	TDK	Stealth Aircraft
Carbon composite materials	Toray	FSX wing
Components		
Charge-coupled devices	Mitsubishi Electric	Missile guidance
Television camera	Hitachi	Remotely piloted vehicle
Subsystems		
Doppler radar (for cars)	Fujitsu	Aircraft guidance
Satellite ground receiver	NEC	Military receiver
Systems		
BK 117 A3 helicopter	KHI	Could be equipped with antitank missiles

systems has grown from 34 percent in 1981 to 40 percent in 1990, and will increase to 43 percent by the end of the century.[26] Richard A. Linder, president of Westinghouse's Defense Electronics Group, suggests that electronics will be even more important for the emerging military technologies of the future, particularly stealth and multispectral systems. He argues that four technologies that will make crucial contributions to military systems are: (1) very high-speed integrated circuits, (2) digital gallium arsenide circuits, (3) microwave monolithic integrated circuits, and (4) mercury cadmium telluride for infrared detectors.[27] Japanese corporations excel in all four of these technologies. "Thanks to the 'electronics-ization' of defense," says Mitsubishi Electric managing director Takeshi Abe, "the stage is finally set for Japan to build weapons even better than those made in the U.S.A."[28]

Japan leads in other important dual-use areas as well. (See Table 2-1.) Japan, for example, leads in advanced industrial ceramics, which can be used to coat aircraft engines or to hermetically insulate missile guidance systems and warheads. Japan's advanced carbon composites and radar technology have enabled it to achieve world leadership in important subsystems for the FSX.[29] The Technical Research and Development Institute (TRDI) has been successful in adapting commercial computer technology for use in cockpit control systems. The Japanese are particularly advanced in the miniaturization of electronic hardware which is so valuable for military aircraft. The Japanese journal *Voice* estimates that Japan leads the United States in thirteen categories of technology of important military use, whereas the United States leads in only seven and the two countries are even in two.[30] According to the DOD, Japan is significantly ahead of the United States in some niches of five of twenty technologies that are critical to national security and "the long-term qualitative superiority of U.S. weapon systems"[31]:

Semiconductor materials and microelectronic circuits*
Software producibility
Parallel computer architectures
Machine intelligence and robotics*
Simulation and modeling
Photonics*
Sensitive radars
Passive sensors
Signal processing
Signature control
Weapon system environment
Data fusion
Computational fluid dynamics
Air-breathing propulsion
Pulsed power
Hypervelocity projectiles
High-energy density materials
Composite materials
Superconductivity*
Biotechnology materials and processes*

*Areas where Japan leads significantly in some niches.

America's primary rival for preeminence in these technologies is no longer the Soviet Union, but Japan.

The U.S. DOD has not overlooked the enormous potential for the military use of Japanese commercial technology. In fact, the Pentagon already depends heavily on Japanese components, particularly memory devices, for its weapon systems. In 1980, the DOD and the JDA established the Systems and Technology Forum to explore avenues for cooperation in military research and development, production, and procurement. In 1983, Prime Minister Yasuhiro Nakasone announced that Japan would make an exception to the country's arms export ban for exports of military technology, but not for military systems, to the United States. In November of that year, Japan and the United States signed notes establishing a Joint Military Technology Commission (JMTC) comprising State and Defense representatives from the American embassy in Tokyo and Japanese representatives from the Defense Agency, MITI, and the Ministry of Foreign Affairs. In December 1985, the two sides followed up with detailed arrangements for the transfer of military technology.[32]

Prior to the FSX project, only three examples of such technology transfers were ever approved, all of which were contrived more for their role as precedents than for any immediate benefit to the United States. Gregg Rubinstein, one of the original architects of the agreement, suggests that the DOD pushed for the exchange procedures not with any expectation of significant transfers in the short term, but in the hope of setting up an apparatus that could bring real payoffs in the 1990s and beyond.[33] The first case involved the guidance and control system for the Toshiba portable "Keiko" surface-to-air missile (SAM), a system heralded in Japan but nonetheless of questionable value to the U.S. military. The JMTC approved the government-to-government transfer at a price of approximately $700,000 in December 1986, but the sale was never made due to the political fallout from the Toshiba Machine incident. In the second case, an industry-to-industry transfer, Ishikawajima-Harima Heavy Industries (IHI) sold shipbuilding technology for tactical auxiliary oil tankers to the Pennsylvania Shipyards of the Military Sealift Command. In the final case, an industry-to-government transfer, IHI sold its expertise to the U.S. Navy's Philadelphia Shipyard for overhauling the U.S. aircraft carrier *Kitty Hawk* under a service-life extension program. Both of the latter two transfers met real needs of the U.S. Navy, but they probably could have been arranged as commercial technology transfers if they had not been such convenient trial cases for the new military technology transfer arrangements.[34] In any case, the DOD has always been more interested in Japanese dual-use technology than in strictly military technology.

Since 1983, the DOD has sent a series of technology assessment teams to Japan to evaluate Japanese technologies of potential military use. The first such team, a Defense Science Board task force that toured Japan in November 1983, cited sixteen primary areas of interest in Japanese technology:

Gallium arsenide devices—microwave, high-speed logic
Microwave integrated circuits
Fiber-optic communications
Millimeter-waves
Submicron lithography
Image recognition
Speech recognition/translation
Artificial intelligence (knowledge-based computer architecture)
Electro-optical devices
Flat displays
Ceramics (for engines, electronics)
Composite materials
High-temperature materials
Rocket propulsion
Computer-aided design
Production technology (including robotics, mechatronics)[35]

The DOD sent a technology team to Japan in July 1984 and April 1985 to look specifically at electro-optics and millimeter/microwave technology, and sent a follow-up team in August 1986.[36] In January 1987, Dr. Clinton Kelly led a Defense Advanced Research Projects Agency (DARPA) mission to look into Japanese manufacturing technology in electronics, heavy machinery, and avionics.[37] Jamieson C. Allen, former director of military R&D exchange at the U.S. embassy in Tokyo, argues that these efforts failed to result in significant transfers of Japanese technology primarily because U.S. defense contractors were largely uninterested in Japanese technology.[38]

In 1988, Japan proposed that the United States and Japan work together on research and development in five areas of military technology. In 1990, the two sides agreed to work together on three of these five areas: ducted pocket technology, for rocket engines; hybrid seeker technology, for missiles; and closed-loop degaussing technology, which makes submarines less susceptible to detection. During the same year, the United States proposed seven more areas of cooperation. The DOD and JDA formed working groups on two of these: fighting vehicle propulsion using ceramic materials, and advanced steel for ships and armored vehicles. As noted previously, the SDIO is handling cooperation in another two of these areas: Josephson–Junction microprocessors and the fabrication of artificial diamonds. The DOD and JDA have not yet decided whether to work on the final three areas: ferro-electric technology, military DRAMs, and opto-electronics and optical components.[39]

Needless to say, the JDA and Japanese defense contractors are also exploring the possible military applications of their commercial technology. The JDA's TRDI has set up a small bureau to gather information on such dual-use technology, and to coordinate the process of directing this technology into military applications. In July 1987, the TRDI reorganized in order to better concentrate on areas of Japan's greatest potential strength such as optics, electronics, and command, control, communications, and intelligence (C^3I). The TRDI and the defense contractors are particularly eager to incorporate some of the hottest new dual-use technology into weapon systems, particularly the FSX. In order to maximize the benefits from their dual-use technology, the Japanese must be able not only to adapt this technology to new uses, but to mix and match a whole range of such dual-use components to produce an integrated weapon system. Those who are most skeptical about Japan's ability to catch up with the United States in military technology stress the difficulty in bridging the gap between producing isolated parts and producing complete systems. The Japanese may have the necessary technology, the argument goes, but they cannot integrate a system.

Japanese Weapon Systems

To those who claim that the Japanese cannot integrate weapon systems, one could simply respond by pointing out that they already do. The Japanese produce military aircraft, warships, tanks, and missiles—some under license, but others on their own. But this response begs two more difficult questions. First, to what extent do the Japanese rely on U.S. technology and U.S. parts? And second, just how good a system can they produce? To better understand Japan's technological capabilities and limitations, we need to disaggregate the know-how required to build a weapon system. As noted in the previous two sections, Japan excels in all areas of basic technology needed to produce a complex weapon system such as a fighter aircraft. Japan may in fact surpass the United States in selected areas such as electronic devices and coating materials.

Moving up one level on the ladder of integration, however, Japan has some decided weaknesses when it comes to large subsytems, particularly jet engines. In fact, even under the plan for domestically developing the FSX, JDA officials were resigned to the fact that they would have to import the engines. Japanese manufacturers are more accomplished at producing other subsystems such as the computer and communications systems for an aircraft cockpit. Japan's greatest weakness comes in the realm of overall technological know-how in areas such as aerodynamics and, of course, system integration. (See Table 2-2.) Japanese contractors lag in these areas primarily because of their inexperience in developing their own weapon systems. They have advanced considerably through the repeated exercise of producing under license, but they will be able to master the subtleties of system integration only through the experience of developing their own new systems or at least co-developing them with foreign producers. It is not surprising, therefore, that these contractors were so determined to develop the FSX indigenously.

The United States–Japan technology gap in weapon systems should not be underestimated. In most weapon systems, the United States is a full generation (five or more years) ahead of Japan. Furthermore, the Japanese defense industry has learned practically everything it knows from its senior partner in the United States. At the same time, however, we should not underestimate Japan's ability to close the gap, given the political will. Japanese defense contractors

Table 2-2 The Technology Necessary to Make an Airplane

Area	Examples
Materials and Components	
Advanced materials	Aluminum composites, ceramics
Electronic parts	Integrated circuits, gyros
General parts	Seats, ventilation system
Structure	Welding, hardening
Subsystems	
Flight control	Automatic pilot, sensors
Operation	Radar, cockpit controls
Propulsion*	Engine, exhaust system
Know-How	
Aerodynamics*	Wing design, computer analysis
Production technology	Laser processing, FMS
Testing & evaluation*	Weather tunnel, simulation
System integration*	Overall design, simulation

*Areas where Japan lags considerably behind the United States.

Source: Compiled by the author, based in part on information from the Japan Aerospace Industry Association.

Table 2-3 Composition of the Defense Budget, Fiscal 1986–91

	1986	1987	1988	1989	1990	1991
Personnel and Provisions	45.1%	43.9%	42.7%	41.2%	40.1%	40.1%
Supplies	54.9	56.1	57.3	58.8	59.9	59.9
Equipment acquisition	26.9	27.5	28.1	28.0	27.4	27.7
Research & development	1.7	1.9	2.0	2.1	2.2	2.3
Facility improvements	1.7	2.0	2.8	2.9	3.2	3.1
Maintenance	14.4	14.2	14.1	15.1	16.1	15.9
Base countermeasures	9.0	9.4	9.2	9.5	9.8	9.7
Other	1.2	1.1	1.1	1.2	1.2	1.2

Source: Japan Defense Agency, *Boei hakusho 1991* (Defense White Paper 1991), p. 261.

have managed to license essential know-how from the United States and to expand their own capabilities to the point where they are now positioned to develop their own weapon systems within a reasonably short period of time. The Japan Defense Agency continues to import those systems that cannot be produced at home, while doing its best with its limited budget to close the technological gap with the United States. The JDA has been remarkably successful at staying not too far behind, with only a very modest investment in military R&D. "Up until now," remarks Sanshiro Hosaka of the TRDI, "we have always been running behind the United States. That's why we have been able to research and develop so efficiently, learning from the Americans' mistakes. And now, all of a sudden, we have some of the best technology in the world."[40] In recent years, the JDA has significantly boosted the portion of the defense budget directed at R&D, reaching 103 billion yen, or 2.3 percent of the defense budget, in fiscal 1991. (See Table 2-3.)

The JDA has consistently sought to decrease its reliance on foreign technology. It is impossible to judge precisely how much of Japan's defense equipment is produced domestically because JDA figures grossly overestimate this percentage by accounting for completed systems only. The JDA counts a system produced in Japan under U.S. license, for example, as 100 percent domestically produced. JDA figures for 1989 show that 90.4 percent of total defense procurement is domestically purchased.[41] Nevertheless, Japan still produces many of its more sophisticated systems under license and it relies on the United States for many of the most important parts. The Japanese contractors themselves estimate that U.S. manufacturers produce 15-40 percent of the total value of Japanese defense production. JDA officials give several reasons for not wanting to rely on foreign producers. First, they say that they want to have the ability to maintain and repair their systems at home. These officials still remember when they had to overhaul fourteen reconnaissance versions of the F-4E that were delivered to Japan in 1977–78. "A Japanese contractor would send out a team of technicians immediately if there were any problem with an aircraft," declares Ken Adachi, chief engineer, Radio Group of the NEC Corporation, and former lieutenant general in the Ground Self-Defense Forces (SDF), "and they probably would not even charge for repairs."[42] Second, they argue that they are able to get better systems at a lower price from the United States when they have the option of domestic production. In essence, however, they simply do not like the idea of having to depend on the United States for military hardware. "It gives me chills to think how much we rely on U.S. parts," laments Yasuo Komoda, manager of Fujitsu System Integration Laboratories's R&D Coordination Office and a former major general in the Ground SDF.[43]

Although the prospects vary between subsectors, there is some evidence that suggests that Japanese defense contractors could become competitive producers of major weapon systems within ten to twenty-five years. For one thing, the TRDI has successfully tapped Japan's commercial technology base in developing several world-class military subsystems, the two most famous of which are ready for use in the FSX project. General Dynamics has shown a keen interest in gaining access to the Japanese technology for producing the aircraft's wings out of carbon composite materials. U.S. producers are able to produce carbon composite wings, but they have to cure the wing's two surfaces and the ribs and spars separately. The TRDI and a consortium of Japanese producers have developed a production process whereby the bottom surface of the wing is "co-cured" with the ribs and spars. This alleviates the need for the huge number of heavy rivets that join the bottom surface to the ribs and spars in other aircraft. The result is a wing that is stronger and 40 percent lighter than a more conventional alternative.[44] The other widely heralded subsytem being developed for the FSX is Mitsubishi Electric's (Melco) active phased-array radar. Melco and the TRDI are substantially ahead of U.S. producers of similar radars in that they have been able to make the radar small enough to put on a fighter aircraft and they have developed an extremely efficient cooling system. They have already produced two prototypes and have tested them on a C-1 aircraft at the TRDI's Gifu test center. The radar, which has more than a thousand "active" radiating elements, boasts ultra-high resolution and unprecedented terrain-mapping capabilities.[45] The *Asahi Shimbun's* economic news desk, in its book on the power of military technology—or "militech power"—points out that the TRDI was only able to make the radar small enough to put on a military aircraft because of civilian industries' development of high-performance gallium arsenide semiconductors and high-density integrated circuits.[46]

Although most JDA officials and the primary contractors would have preferred to develop the FSX indigenously, they still hope to use the co-development experience to try out their best dual-use technology and to improve their skills in system integration. The TRDI has been relatively successful in developing those technologies that the United States will not transfer. The TRDI, for example, has developed its own control configured vehicle (CCV) technology. CCV aircraft are inherently unstable because they have smaller "canard" wings, but they are much more agile than conventional aircraft. They can "slide" horizontally where a traditional aircraft would have to make a banking turn. The FSX will use a new digital fly-by-wire system to continuously monitor flight parameters and instantly readjust in order to maintain balance. The

TRDI has already tested CCV technology on a remodeled T-2 trainer. TRDI officials have been forced to develop their own source codes (software) for the FSX's flight control system because the United States decided not to transfer this technology in response to the congressional uproar of 1989. The TRDI and MHI may not be able to develop the FSX into an airplane that can challenge the U.S. aircraft that are now being developed, but they will gain invaluable experience in the process. "We would like to catch up with the generation after the FSX," declares Sakichiro Ono, chief executive director of the Japan Defense Industry Association and former major general of the Self-Defense Forces.[47]

The TRDI has been remarkably successful in indigenously developing some of the smaller weapon systems, particularly air-to-surface and surface-to-surface missiles (ASMs and SSMs). Mitsubishi Heavy Industries started developing the ASM-1 (Type 80) missile in 1973, and began production in 1980. The F-1 and other fighter aircraft now carry the 50-kilometer range, Mach 1 speed missile for attacks on surface ships. The missile uses inertial guidance in midcourse and active radar homing in its terminal phase.[48] MHI has been lauded for completing development within budget and on schedule, and for producing a missile that has achieved exceptional hit rates in field tests. In 1979, MHI began development of a surface-to-surface missile, the SSM-1, based on the ASM-1. MHI designed the missile for the Ground SDF with a range of 150 kilometers so that it can be launched from points approximately 100 kilometers inland and still strike enemy ships well offshore. The missile is launched by rocket off a special MHI truck. The turbojet-powered cruise missile then uses inertial guidance in its overland phase and part of its oversea phase, but switches to active radar homing as it skims over the water toward its target.[49] The Ground SDF tested the missile at Point Mugu, California, in 1987, and MHI executives claim that American observers were astounded by its superb hit rate.[50] The *Asahi Shimbun* reported that nine out of ten firings either hit the target or landed close enough to be considered "hits."[51]

With the success of the SSM-1, the TRDI and MHI are now working on three more ASM-1 derivatives: an XSSM-1B ship-to-ship missile, an XASM-1C air-to-surface missile for use on P3C aircraft, and an XASM-2 air-to-surface missile for use on F-1 and FSX aircraft. On the XASM-2, the TRDI is using a highly advanced infrared image homing system. The TRDI is also working on an XAAM-3 air-to-air dogfight missile for the Air SDF as a successor to the U.S. AIM-9 (Sidewinder) series. In 1989, the institute began developing an improved version of the short-range surface-to-air missile, the "Tan-SAM." The new Tan-SAM will have independent active radar homing and infrared image homing capabilities.[52] The TRDI is also considering indigenous development of a midrange surface-to-air missile as a successor to the American "Hawk," although it has not made any official announcement of its plans.[53] "We have caught up with the Americans in missile technology," claims one TRDI bureaucrat, "but we have only been able to do so because of the high-performance semiconductors, high-density integrated circuits, quality control, and microprocessors that have come from Japan's industrial technology base."[54] (See Table 2-4.)

Table 2-4 Some Weapon Systems Currently Being Developed by the TRDI

Year R&D Started	System
Aircraft	
1983	(Shipboard) antisubmarine helicopter
1988	Fighter support aircraft (FSX)
1991	Unmanned aerial observation vehicle
Guided Weapons	
1988	Air-to-surface missile (XASM-2)
1989	Improved surface-to-air missile "Tan-SAM"
1990	Anti-ship/anti-tank missile (XATM-4)
Electronics and Communications	
1988	Improved division communications system
1990	Shipboard fire control system
1991	Expendable electronic jammer

Source: Japan Defense Agency. *Boei hakusho 1991* (Defense White Paper 1991), p. 297.

Japan's success in integrating sophisticated space systems also supports the assertion that the Japanese might meet with success in the integration of military systems, if they were willing to make the requisite investment. The National Space Development Agency (NASDA) has gradually decreased its technological reliance on the United States. NASDA's early launch vehicles, the N-1 and the N-2, were approximately 60 percent domestically produced. They were first used in 1975 and 1988, respectively. The H-1, which was first used in 1986, was more than 80 percent domestically produced. The H-1 marked an important transition in Japan's space development, as Japan developed its own cryogenic propulsion technology for the second stage because the United States did not want to relinquish its own technology. NASDA has refused all assistance in developing the H-2, which will be able to launch 4,400 pounds, more than a U.S. Air Force Titan 34 D. The H-2 was expected to be ready for launch by late 1992.[55] NASDA has been equally successful with satellites. For example, the BS-1, launched in 1978, was 90 percent U.S.-produced. The BS-2A and BS-2B, launched in 1984 and 1986, respectively, were 70 percent U.S.-made. And the BS-3A, launched in 1990, was almost entirely domestically produced.[56] The Space Activities Commission's (SAC) Long-Term Policy Council produced a report on May 26, 1987, declaring that Japan is aiming for no less than a "central role" in the global space market by the beginning of the twenty-first century. The report states that Japan will complete the Engineering Test Satellite-6 and the H-2 booster without foreign assistance, and will launch a Japan Experiment Module by the mid-1990s. From the late 1990s into the beginning of the new century, Japan will develop an operating space station and will move on to manned space activity—moon and planet exploration.[57]

One weapon system that the Japanese have chosen not to produce is the nuclear weapon. The choice is primarily, of course, a political one and not one dictated by a lack of technological capability. Japan has a small amount of low-quality uranium in Okayama, but this would not be enough for a significant nuclear arsenal. The Chinese have pointed to the fact that this uranium has never been used for energy purposes as evidence that Japan intends to go nuclear, but the Japanese contend that it simply would not be economical to try to dig it up. In order to divert its uranium from energy use, Japan would have to choose between trying to fool the International Atomic Energy Agency inspectors—a formidable task—or abrogating the Non-Proliferation Treaty (NPT).[58] Japan has a "pilot scale" enrichment plant at Ningyo Pass that produces uranium enriched to 3 percent (of the U-235 isotope) but that could be relatively easily converted to produce weapons grade, 93 percent enriched uranium. Yatsuhiro Nakagawa, a professor of international politics at Tsukuba University and Japan's self-proclaimed "lone advocate" of nuclear armament for Japan, estimates that Japan could be producing weapons-grade uranium within two years if it chose to do so.[59] Alternatively, Japan could use plutonium for nuclear weapons. Japan has large stockpiles of plutonium from spent fuel that was reprocessed in Europe. Under an April 1988 agreement with the United States, for the next thirty years Japan no longer needs to request U.S. permission on a case-by-case basis to send its used nuclear fuel to Europe for reprocessing. Japan also has a "pilot scale" reprocessing (plutonium extraction) plant of its own and is planning to build a larger one within the next decade. Japan could adapt its nuclear reactors in order to produce weapons-grade plutonium, but this would be considerably more difficult and more costly than converting uranium enrichment plants so they can produce weapons-grade uranium.[60] Japan has a variety of delivery systems—tactical aircraft, missiles, and long-range artillery—that could be equipped to deliver nuclear warheads.

In the long term, a reemergence of the Japanese defense industry could have some rather ominous implications. One 1982 report estimated that Japan eventually would capture 60 percent of the market for naval ships, 40 percent of military electronics, 46 percent of military automobiles, and up to 30 percent of the aerospace market.[61] These estimates may be unrealistic, but nonetheless there are indications that Japanese companies could be successful in exporting military equipment. Japan's "reliability" advantage already extends into the military sector. Fujitsu's Yasuo Komoda complains that the quality of the U.S.-made chips he buys for military requirements is abysmal: "Sometimes only 10 percent work."[62] The Westinghouse APQ 120 radar for the F-4 fighter reportedly lasts an average of eight hours before failure, whereas the Mitsubishi Electric equivalent for the F-4EJ lasts an average of forty hours.[63] And the readiness rate for the Japanese-made F-15J is higher than that for the U.S.-made F-15.[64] Although the Japanese weapon systems may have been used under less demanding conditions than the American systems, they have still fared astonishingly well. In the future, Japanese producers may be able to use flexible production systems to bring the low-cost advantages of mass production to the specialized production of advanced military systems. They also may be able to achieve shorter product cycles, which will enable them to incorporate more advanced commercial components into their weapon systems.[65]

Japan's technological strengths could play right into the needs of the warfare of the future. In the twenty-first century, much of America's and Europe's present hardware may be obsolete. Meanwhile, laser weapons and robot soldiers could become a reality.[66] The U.S. and European military establishments would probably not be caught entirely by surprise with such developments, but they would certainly not enjoy the technological lead over Japan that they now have. To the extent that new technology makes present technology obsolete, it will be that much easier for Japan to catch up. Robert J. Art illustrates this point in his discussion of the British development of the dreadnought in 1906. The dreadnought, with its greater power and range, made all other battleships obsolete. By developing the dreadnought, however, the British inadvertently wiped out their own significant lead in pre-dreadnought battleships.[67]

The Political Context

Ultimately, the future of the Japanese defense industry depends on the course of Japanese defense policy. Many commentators, particularly Japanese ones, have argued that Japanese military expansion is strictly constrained by domestic political forces. Although this was certainly true in the 1950s and the 1960s, it was less so in the 1980s—and is even less so in the 1990s. Over the course of the 1970s, the Yoshida "consensus" on national strategy gave way to one that points in a very different direction. The Yoshida consensus refers to a policy originally adopted under the administration of Prime Minister Shigeru Yoshida (1948–54), under which Japan would control internal security while depending primarily on the United States for protection from external threats. According to Article Nine of the postwar constitution, "war potential will never be maintained." The consensus was gradually embodied in a series of explicit constraints on defense expansion, including the three nonnuclear principles, the ban on arms exports, and the ceiling on defense spending of 1 percent of GNP.

A number of developments throughout the 1970s served to erode this consensus. The simultaneous rise of Japan as an economic power and the relative decline of the United States provided the most basic impetus behind the transition. The Japanese became increasingly aware of this shift in world power as the United States suspended the convertability of the dollar into gold in 1971 and withdrew from Vietnam in 1975, and as Japan recovered remarkably well from the oil shocks of 1973 and 1978. Japanese leaders began at the same time to question American protection and to reconsider their country's role in the world. In the late 1970s, a major Soviet build-up in the Far East moved many of these leaders to reexamine their benign view of the Soviet threat. During the same period, officials in the Carter administration (1977–81) and U.S. congressmen began to call for greater Japanese efforts on defense. This "foreign pressure" had a tremendous impact on Japanese attitudes, for Japanese leaders are acutely aware of their dependence on the United States in the economic as well as the security realm.[68]

The new mainstream policy did not kill off those within the Liberal Democratic Party (LDP) and within the ministries who favor strict limits on defense, but it did strike a forceful blow against them. By the late 1970s, the majority of Diet members had come to accept the "realist" position that favors gradual defense expansion, essential cooperation with U.S. requests for sharing a greater portion of the defense burden, and an effort to make Japan's defense more closely designed to meet specific military threats. In 1980, a *Nihon Keizai Shimbun* poll showed that 78.6 percent of the LDP Lower House Diet members advocated expansion of the Self Defense Forces, whereas only a handful were opposed. The Soviet Union was viewed as posing a "major threat" by 41.4 percent, and a "potential threat" by 50.5 percent. The arms export ban was opposed by 46.4 percent, and 37.7 percent favored revision of the Peace Constitution.[69] A group of Dietmen known as the Defense Tribe (*boei-zoku*) successfully lobbied to make defense a top priority item within the budget in the 1980s, and the 1 percent of GNP ceiling was surpassed in 1987. (See Table 2-5.)

With the new direction in policy, a new group of "military realists" tried to come up with a strategy to match Japan's new role. Previously, Japanese defense policy was designed simply to meet a vaguely defined "limited small-scale" attack. Beginning in 1980, Defense White Papers referred to the Soviet

Table 2-5 The Defense Budget, 1955–92

Year	Defense Budget (billion yen)	% Increase over Previous Year	As % of GNP	As % of Total Budget
1955	134.9	3.3%	1.78%	13.61%
.....				
1965	301.4	9.6	1.07	8.24
.....				
1975	1327.3	21.4	0.84	6.23
1976	1512.4	13.9	0.90	6.22
1977	1690.6	11.8	0.88	5.93
1978	1901.0	12.4	0.90	5.54
1979	2094.5	10.2	0.90	5.43
1980	2230.2	6.5	0.90	5.24
1981	2400.0	7.6	0.91	5.13
1982	2586.1	7.8	0.93	5.21
1983	2754.2	6.5	0.98	5.47
1984	2934.6	6.6	0.99	5.80
1985	3137.1	6.9	0.997	5.98
1986	3343.5	6.6	0.993	6.18
1987	3517.4	5.2	1.004	6.50
1988	3700.3	5.2	1.013	6.53
1989	3919.8	5.9	1.006	6.49
1990	4159.7	6.1	0.997	6.28
1991	4386.0	5.45	0.954	6.23
1992	4551.8	3.8		

*1992 figures are estimated (*JEI Report* 1B, January 10, 1992).

Source: Japan Defense Agency, *Boei hakusho 1991* (Tokyo: JDA, 1991, p. 262).

Table 2-6 Japan's Top Ten Defense Contractors, Fiscal 1990

Company	Sales (billion yen)	Defense contracts as percent of total sales
1. Mitsubishi Heavy Industries	440.8	18.9%
2. Kawasaki Heavy Industries	146.5	16.4
3. Mitsubishi Electric	100.3	3.9
4. Ishikawajima-Harima Heavy Industries	78.6	10.7
5. Toshiba	59.9	1.9
6. NEC Corporation	54.5	1.8
7. Nihon Seikosho	34.8	26.1
8. Komatsu	22.4	3.3
9. Fuji Heavy Industries	21.6	2.9
10. Hitachi	20.1	0.5

Source: Federation of Economic Organizations (Keidanren).

Union by name as the principal military threat and have justified force levels by the need to defend Japan against a possible Soviet attack. In a May 1981 meeting with U.S. President Ronald Reagan, Prime Minister Zenko Suzuki extended Japanese commitments by agreeing that Japan would defend its own sea lanes for a distance of 1,000 nautical miles out. Japanese acceptance of this mission implies a need for much greater reconnaissance and anti-submarine warfare capabilities. U.S. and Japanese officials have exchanged information on defense planning more freely since the establishment of the "Guidelines for U.S.–Japan Defense Cooperation" in 1978, and U.S. and Japanese forces have performed combined military exercises involving all branches of the services since 1986.[70]

In the future, the Japanese defense industry may become an important political force in favor of military expansion. Defense production accounts for a meager 0.5 percent of total production for Japanese industry, and even for the largest contractors defense production accounts for only a small portion of sales.[71] Yet at the same time, Japan's top defense contractors are also some of its largest and most powerful corporations. (See Table 2-6.) These corporations expend more resources and more political capital on their defense business than might be justified by defense sales alone. "Defense may only account for three percent of our business," says Kunio Saito, general manager of NEC's 1st Defense Sales Division, "but it certainly takes up more than three percent of our energy."[72] Defense requires more political effort because it is a political business. Defense contractors have only one client, the Japan Defense Agency, so the incentives to lobby are great. Companies see defense as a secure business, insulated from the pitfalls of the business cycle, and they expect the defense sector to continue to grow steadily. "By the year 2000, we are confident that our sales will grow to a level warranting the kind of investment we are making today," declares Yotaro Iida, president of Mitsubishi Heavy Industries.[73] Even more importantly, Japanese firms see involvement in the defense business as an imperative so as not to fall behind in the high-tech race. They envision commercial spin-offs from defense production, and they fear that they may miss out if they are not at least peripherally involved in the defense business. They see the defense industry as one that may drive innovations in other areas, such as electronic components.[74] "We are being challenged by the NICs in traditional consumer markets," explains Fujitsu's Komoda. "We have to go value-added, and all that is left is space and defense."[75]

Defense contractors use three primary channels to lobby the government. First, they hire retired Self-Defense Forces officers to serve as intermediaries with the JDA and the forces. Top bureaucrats have long had a tradition of "descending from heaven" (*amakudari*) into prominent roles in private industry after retirement at age fifty-five or sixty, and military officers have followed this same practice. A defense contractor, it is said, should have at least one military "old boy" for every 20 billion yen in annual defense sales. Second, they work through the industry associations, the most important of which is the Japan Defense Industry Association (*nihon boei sobi kogyokai*). Established in 1951 as the Japan Ordnance Association (*nihon heiki kogyokai*), the association was renamed and reorganized in September 1988. It is now an incorporated association officially affiliated with both the JDA and MITI.[76] Finally, the contractors join forces in the Defense Production Committee of the Federation of Economic Organizations (Keidanren), Japan's most powerful business organization and a primary source of funding for the ruling LDP. In 1989, Keidanren came out with a public position paper on the Mid-Term Defense Plan for 1991–95, demanding that R&D spending be doubled and that the domestic content of JDA procurement be raised further.

In the wake of the U.S. response to the FSX agreement in 1989, a number of prominent LDP Dietmen and JDA officials have called for a renewed push to decrease reliance on U.S. military technology. Popular LDP Dietman Shintaro Ishihara, for one, argues that Japan should build the FSX on its own. The fact that Japan gave in to the U.S. demand to co-develop the plane, he stresses, confirms that Japanese leaders are still excessively deferential to U.S. power.[77] Even JDA officials suggest that they may not want to do any more co-development with the United States. They estimate that they have lost at least one year because of U.S. delay in approving the agreement. "Frankly," proclaims one senior JDA official, "we are no longer open-minded about co-development. The United States caused the delay, but we have paid the price."[78]

The opposition parties, particularly the Japan Socialist Party (JSP) and the Japanese Communist Party (JCP), have taken the lead in criticizing LDP policies on defense. To date, their protests have not had much effect on the ruling LDP, which has enjoyed a stable majority in both houses of the Diet. The opposition parties gained a bit more of a voice with their victory over the LDP in the July 1989 Upper House (House of Councillors) elections, but the LDP managed to stay in power by retaining its majority in the more powerful Lower House (House of Representatives) after the February 1990 elections.[79] According to the Constitution, the Lower House can pass the budget without the approval of the Upper House, but it needs a two-thirds majority to override the Upper House on ordinary bills.[80] In recent years, the JSP and the Komeito have become much more moderate in their opposition to the LDP's defense policy, and some members of the Democratic Socialist Party have become downright hawkish. The JCP has stuck to its principles, but it has found itself increasingly isolated.

The present "realist" consensus is not likely to give way to a more assertive defense posture for Japan unless international developments once again transform the national consensus on strategy, as they did in the 1970s. In general, Japanese defense policy is more responsive to the state of the U.S.–Japan relationship than to the actual military threat to Japan. Japanese leaders' desire to increase their country's military role will grow to the extent that they perceive a decline in the U.S. capability or intention to protect Japan. In short, if the United States cannot or will not protect Japan, Japan will defend itself. Some U.S. policymakers may welcome such a development, but if Japan does choose to take over full responsibility for defense, the United States will lose control over Japanese force levels and military doctrine, and the United States will sacrifice much of its leverage over Japanese foreign policy. Japanese leaders will be particularly sensitive to U.S. economic weakness, U.S. military weakness, U.S. pressure on Japan to do more for itself, and a partial or complete withdrawal of U.S. forces from Japan.

JDA officials are particularly sensitive to America's unwillingness to share its latest military technology. They claim that the U.S. DOD has been much less forthcoming with defense information and technology since 1980.[81] The defense industry has been particularly attuned to the efforts of U.S. companies to "black-box" military technology sold to Japan; that is, to package it so that Japanese firms will not be able to copy it. Although U.S. leaders see this as nothing more than good business and the protection of America's own security interests, the Japanese perceive this as a sign of U.S. distrust and an important reason why Japan should build up its own military production capability.

The Japanese media used the thirtieth anniversary of the United States–Japan Security Treaty in 1990 as an opportunity to reevaluate Japan's dependence on the United States in a variety of special feature articles. With the end of bipolarity, some analysts suggest the Japanese should seriously question the American commitment to defend Japan. An *Asahi Shimbun* poll reported that only 31 percent of Japanese said that "Japan should continue to depend on the United States," whereas 40 percent argued that "Japan should build up an independent defense system."[82]

The Japanese desire for greater autonomy has only grown as United States–Japan trade friction gets worse. Japanese government officials are already bitter about what they see as unjustified "Japan-bashing" by the United States. They feel that a lack of diligence and poor management has reduced the competitiveness of U.S. products, and that U.S. critics of Japan are simply using Japan as a scapegoat for their own woes.

Although Japanese military strategists finally found a clearly identifiable enemy in the Soviet Union in the 1980s, they may have lost it again in the 1990s. JDA officials argue in the 1990 Defense White Paper that Soviet President Mikhail Gorbachev's reforms had not altered the basic military threat posed by the Soviet Union. Former JDA Director-General Koichi Kato (currently an LDP member of the House of Representatives) even suggests that a decrease in the Soviet nuclear threat implies an increase in the conventional threat, which is what Japan is most concerned with in the first place.[83] Most LDP leaders, however, feel that they are compelled to respond to the end of the Cold War by slowing down the expansion of the defense budget. In December 1990, the Cabinet compromised on a total outlay of 22.7 trillion yen for the five-year period of the midterm defense plan for 1991–96. This figure implies annual spending growth of only about 2.9 percent.[84]

In the future, Japanese leaders may be more sensitive to military threats from other countries in Asia. They continue to closely monitor events in China, and they are particularly concerned with the political instability that plagues South Korea and the Philippines. Japan will react strongly to heightened tension between the Koreas or to a weakening of the U.S. commitment to South Korea. In addition, Japanese leaders may be more unsettled than relieved if Korea becomes reunified.

The 1990–91 Gulf War between Iraq and U.S.-led multinational forces provided a new stimulus to the Japanese debate over defense policy. Prime Minister Toshiki Kaifu, with considerable prodding from the United States, tried to push through a bill that would have sent Japanese nonmilitary personnel to support the multinational forces in the Persian Gulf region. Kaifu and LDP Secretary General Ichiro Ozawa were unable to unify their own party behind the plan, and thus Japan ended up contributing money—about $9 billion of it— but not manpower to the war effort. The debate over this bill reopened discussion over one of the most powerful constraints on Japan's military role: the ban on sending military forces overseas. The war also revitalized the perennial national debate over Japan's role in the world. Some Japanese leaders were ashamed that their country was not able to act more decisively, whereas others simply accused the United States of making Japan pay for its own follies.

The Power behind Technology

Japanese leaders may not ever decide to pursue a truly independent defense strategy. But even if they do not, Japan will have more influence over international security relations in the years to come by virtue of its economic and technological strengths alone. Japanese technological leadership gives Japan more power within the international system in three respects. First, Japan will gain leverage in its relationship with the United States as the DOD relies more on Japanese technology and the JDA relies less on U.S. technology. Martin Libicki and his colleagues suggest that U.S. dependence on foreign components and technology impairs the country's "surge capability"—its ability to accelerate the production, maintenance, and repair of critical items during a conflict. In addition, this dependence creates "technology base vulnerability" because the

United States may lose access to the most advanced technology for the development and production of weapons.[85] The Defense Science Board Task Force on Semiconductor Dependency suggests that DOD reliance on Japanese semiconductors seriously threatens U.S. national security interests because the United States cannot count on maintaining a technological lead if it does not control the production of crucial electronic components.[86]

Those who downplay U.S. dependence argue that the United States could produce just about anything that the Japanese can, albeit at a higher cost. This begs three more important questions: At what cost? How quickly? And most importantly, just how good would the U.S. substitute be? Cost is a factor, even in military affairs. The United States will not suffer seriously even if it has to pay $1 million for a crucial semiconductor. It will lose out, however, if it has to pay more for a whole variety of components ranging from semiconductors to costly subsystems. In addition, there is a difference between being able to produce something eventually and being able to produce something today. In a crisis situation, the U.S. military may not be able to wait around for domestic producers to come up with an item that previously had been "made in Japan." The U.S. military will not be able to escape its dependence if American products, at any cost, are not as reliable as the Japanese ones. A domestic substitute will do more harm than good if it does not function properly. Furthermore, even if U.S. manufacturers have the ability to produce many of the components now imported from Japan, they may lose further ground if they are not actually producing them. Through a gradual process of product improvement and production innovation, the Japanese firms that manufacture these components may come up with advances that their idle American competitors will not be able to emulate. Semiconductor dependence is particularly a problem because Japan's top semiconductor manufacturers are also Japan's top computer manufacturers. NEC, for example, might find it to be in its interest to withhold the technology for its most advanced semiconductors so that it would have an advantage in competition with U.S. computer makers. Alternatively, Japanese producers might be more interested in the more lucrative commercial market, and therefore might be unwilling to produce their parts to military specification.[87]

How will this dependence translate into political leverage? Japan could use the threat of halting exports at a crucial period to gain its own political goals. Libicki suggests that this could only work once because the United States would quickly move to compensate for its dependence on Japan through research and development.[88] The cycle of offering and selectively denying the United States advanced Japanese technology could, however, develop into an ongoing process. The United States would lose out if Japan began to refuse to export the best of its advanced technology, or if Japan used this as a threat to gain U.S. concessions in other areas. As the well-known Japanese commentator Hajime Karatsu puts it: "If Japan stopped exporting semiconductors, the United States would be turned upside down. This gives Japan an extraordinary amount of bargaining power."[89] In reality, of course, Japan is not likely to stop exporting semiconductors to the United States. U.S. dependence simply means that Japanese officials and businessmen have more leverage in their dealings with their American counterparts. In other words, the balance of power within the bilateral relationship is evening out.

Japanese leaders also feel that they have achieved a more balanced relationship with the United States because they are less dependent on U.S. military technology. The Japanese are confident that they can develop the military technology they need with or without U.S. support. Japanese contractors and the JDA hoped to develop the FSX indigenously in part because they wanted to prove they could do it on their own. They began R&D work years before the 1987 procurement decision in order to prove that they could do so. Now they can use the U.S. renegotiation of the original memorandum of understanding for the FSX as a pretext for developing future weapon systems independently. Even if they do continue to work with U.S. defense contractors, the Japanese feel that they gain a more even hand in a wide array of negotiations with the United States by maintaining a credible threat of going it alone. "Thanks to our military technology base," notes one senior JDA official, "we are bargaining from a stronger position."[90]

Second, Japan now has the ability to play a pivotal role in the global race for superiority in military technology. Although the United States–Soviet rivalry for technological superiority is no longer as intense as it once was, U.S. military strategy still relies on a substantial technological lead over potential and real adversaries, whether they be Soviets or Iraqis. Pentagon planners are all too aware that Japan can help the United States to keep this technological edge, and that is why they have pursued Japanese military and dual-use technology transfers so vigorously. The combination of U.S. strength in basic research with Japanese prowess in applied research, and U.S. sophistication with Japanese reliability, would be unbeatable. At some point, of course, Japanese officials could threaten to withhold their cooperation if they were dissatisfied with the direction of U.S. foreign policy.

Conversely, even occasional technology exports from Japan to U.S. adversaries could undermine the most valiant of American efforts to retain a technological edge. In the wake of the Toshiba Machine affair, the MITI has expanded its corps of technology export control inspectors from 15 to 100 and has established some of the most severe penalties for export violations among U.S. allies. In April 1988 the Japanese government agreed to protect U.S. military technology with registered, classified patents.[91] Nevertheless, most of Japan's best dual-use technology is still readily available, and Japanese companies remain unaccustomed to security controls. Robert L. Mullen, assistant deputy undersecretary of defense for trade security policy, suggests that the problem will intensify by the mid-1990s, after Japan has completed the H-2 rocket launch vehicle. By that time, Japan will have developed a whole array of important components for use in space. Japan may be able to export these as commercial-use products, but the purchasers are more likely to be interested in them for their military uses.[92]

Third, Japan's advances in commercial technology give it the ability to become a major military power in its own right by the early twenty-first century. Japan already has the economic resources to become a military superpower. Japan has enough money to buy itself a world-class defense establishment if it wants to. Japanese GNP for 1990 reached 437.6 trillion yen.[93] Japanese GNP growth has outpaced that of the other major industrialized countries throughout most of the postwar period, and it can be expected to continue at a rate of 3 to 5 percent per year in the near-term future. David Denoon has estimated that Japan could double defense spending without any substantial negative impact on GNP growth. "Within anticipated ranges," he concludes, "limitations on defense spending are political and not due to economic constraints."[94]

Japan also has the human resources to develop a competitive defense industry and to manage a powerful military establishment. Japanese literacy rates are among the highest in the world, and the quality of Japanese education, particularly at the secondary level, is the best in the world. In 1988, Japan had more than 5 million students specializing in science and engineering in high school, more than 2 million in college, and 80,000 in postgraduate courses.[95] Furthermore, the Japanese government has been extremely successful at directing its economic and human resources into productive uses. If Japan were to militarize, its strong centralized state would serve it well. The Japanese government would be well poised to coordinate national research efforts and to stimulate private-sector investment. In addition, the government would have the ability to rapidly and efficiently divert resources from the civilian to the military sector.

Japan's technological leadership provides the final link in giving the country the potential to become a military power. Japan's technology base is unsurpassed even by the United States, and it leads the world in a number of critical dual-use technologies. Japanese producers have gone from a position of clear inferiority to superiority in a whole range of sectors within a remarkably short period of time. Could they do the same in the military sector? The skeptics argue that the technology necessary to produce a fighter plane differs fundamentally from that necessary to produce a Sony Walkman or a Toyota Corolla, and of course they are right. Yet Japanese producers already have some experience in integrating complex military systems, and they have had striking successes in producing some of the smaller weapon systems. Nevertheless, the defense industry would need substantial support from the government before it could make the enormous investment necessary to become truly competitive in a business such as military aircraft production. Without some major changes in the political climate, the government is not likely to offer this kind of full-fledged support.

Even if Japanese leaders never choose to exercise the country's military potential, the mere fact that Japan can become a military superpower will affect international relations in the years to come. Russia and China will grow more concerned about Japan's technological superiority, and they are likely to be more cautious about making any move that could push Japan to accelerate rearmament. The smaller countries of Asia may look more to Japan than to the United States for capital, technology, and perhaps even military protection. They will do so not out of any love for the Japanese, but rather from a recognition that Japan has replaced the United States as the dominant economic power in the region. The United States will also have to consider the costs and benefits of a more powerful Japan, and may be willing to make concessions to Japan in order to ensure that Japan remains a steadfast ally. Whether it is a military superpower or not, Japan at the turn of the century will wield considerable influence in global power politics. Japan will be a "great power" in the international system, although it may be a great power of a new and different kind.

3

Europe's Emergence as a Global Protagonist¹

WAYNE SANDHOLTZ AND JOHN ZYSMAN

Europe is in metamorphosis, reshaping itself and its role in the world. The bargain that is the European Community is being recast by its present members through a process loosely labeled "1992." But more profoundly, Europe is redefining itself—its borders, its character, and its purpose. Central—once labeled Eastern—Europe is being reintegrated into the West. As one might expect, Europe's place in the global system is shifting as well. However labeled, all these changes are themselves a product of shifts in the international system that are altering the options and constraints on European governments.

Dramatic changes in the power and purposes of allies and rivals have provoked this European transformation. The first phase of this transformation, we argue, was sparked by the relative decline of the United States and the emergence of Japan as an economic power; the European Community responded with renewed movement toward a single market. In the second phase, the retreat of Soviet power from Central Europe probably accelerated the discussion of monetary union as a means of addressing the place of Germany in Europe. We must distinguish between the domains of trade and finance (where common European purposes, institutions, and commitments are emerging) and the domains of politics and security (where purposes and institutions for pursuing them are by no means clear). Europe has not yet defined itself as a protagonist in the international system, and may never do so. But a redefined and recast Europe could have an interest in managed multilateralism or regional autonomy. The character of international arrangements that Europe finds congenial will emerge through the process of Europe's transformation.

Under the banner of "1992," the European Community is putting in place a series of political and business bargains that will recast, if not unify, the European market. This initiative is a disjunction, a dramatic new start, rather than the fulfillment of the original effort to construct Europe. The removal of all barriers to the movement of persons, capital, and goods among the twelve member states (the formal goal of the 1992 process) is expected to increase economies of scale and decrease transaction costs. But these one-time economic benefits do not capture the full range of purposes and consequences of 1992.[2] Dynamic effects will emerge in the form of restructured competition and changed expectations. 1992 is a vision as much as a program—a vision of Europe's place in the world. This vision is already producing a new awareness of European strengths and a seemingly sudden assertion of the will to exploit those strengths in competition with the United States and Japan. It is affecting companies as well as governments. A senior executive of Fiat declared, "The final goal of the European 'dream' is to transform Europe into an integrated economic continent with its specific role, weight and responsibility on the international scenario vis-à-vis the U.S. and Japan."[3]

We propose that changes in the distribution of economic power (crudely put, relative American decline and Japanese ascent) triggered the 1992 process. Just as important, European elites perceive that these changes in the international setting require that they rethink their roles and interests in the world. The United States is no longer the unique source of forefront technologies; in crucial electronics sectors, for example, Japanese firms lead the world. Moreover, Japanese innovations in organizing production and in manufacturing technologies mean that the United States is no longer automatically the most attractive model of industrial development. In monetary affairs, some Europeans argue that Frankfurt and Tokyo, not Washington, are now in control. These shifts in relative technological, industrial, and economic capabilities are forcing Europeans to rethink their economic goals and interests as well as the means appropriate for achieving them. American coattails, it seems, are not a safe place when the giant falters and threatens to sit down. American leadership in the war against Iraq might seem a harbinger of renewed American initiative. But the war also showed that military force is the sole area in which the United States is clearly preeminent; President Bush had to pass the hat among U.S. allies to pay for the effort.

Although economic changes have triggered the 1992 process, security issues are shaping its outcomes. The European economic relationship with the United States has been embedded in a security bargain that is being reevaluated. This is not the first reassessment of the alliance, but this time it takes place against the collapse of the Soviet Union. We need not look deeply into the security issues to understand the origins of the 1992 movement, though some believe that the nuclear horsetrading at Reykjavik accelerated the 1992 process.[4] Now, however, the economic and security discussions are beginning to shape each other.

The First Phase

1992: Recasting the European Bargain

The central proposition of this chapter is that structural change was a necessary, though not a sufficient, condition for the renewal of the European project. The creation of the European Community in the 1950s as well as the renewal in the 1980s can be fruitfully analyzed as a hierarchy of bargains. In response to structural changes in the international system, fundamental bargains are struck embodying basic objectives; subsidiary bargains are required to implement these objectives. Europe was created as and remains a bargain among governments. But a broader group of elites helps generate the ideas and vision, sets the context for bargaining, and pushes the process forward.

THE ORIGINAL MOVEMENT

The original European movement can be seen in terms of the following framework.[5] The integration movement was triggered by the wrenching structural changes brought about by World War II; after the war, Europe was no longer the center of the international system, but rather a frontier and cushion between the two new superpowers.[6] Political entrepreneurship came initially from the group surrounding Robert Schuman and Jean Monnet. The early advocates of integration succeeded in mobilizing a transnational coalition supportive of integration; the core of that coalition eventually included the Christian Democratic parties of the six original members plus many of the Socialist parties.[7]

The fundamental objectives of the bargains underlying the European Coal and Steel Community (ECSC) and the expanded European Community were primarily two: (1) the binding of German industry to the rest of Europe so as to make another war impossible and (2) the restarting of economic growth in the region. These objectives may have been largely implicit, but they were carried out by means of a number of implementing bargains that were agreed upon over the years. The chief implementing bargains after the ECSC included the Common Market, the Common Agricultural Program, the regional development funds,[8] and, most recently, the European Monetary System (EMS).

The fundamental external bargain made in establishing the Community was with the United States; it called for (certainly as remembered now in the United States) national treatment for the subsidiaries of foreign firms in the Common Market. That is, foreign (principally American) firms that set up in the Community could operate as if they were European. American policymakers were willing to tolerate the discrimination and potential trade diversion of a united Europe because the internal bargain of the EEC would contribute to foreign policy objectives. Not only was part of Germany tied to the West, but sustained economic growth promised political stability. All of this was framed by the security ties, seen as necessary on both sides of the Atlantic to counter the Soviet Union.

The European bargains—internal and external—were made at the moment of American political and economic domination. A bipolar security world and an American-directed Western economy set the context in which the European bargain appeared necessary. Many expected the original Community to generate ever more extensive integration. But the pressures for spillover were not that great. Economics could not drive political integration. The building of nation-states remains a matter of political projects. Padoa-Schioppa has put it simply and well: "The cement of a political community is provided by indivisible public goods such as 'defence and security.' The cement of an economic community inevitably lies in the economic benefits it confers upon its members."[9] The basic political objectives sought by the original internal bargain had been achieved: The threat of Germany was diminished and growth had been ignited. When problems arose from the initial integrative steps, the instruments of national policy sufficed to deal with them. Indeed, the Community could accommodate quite distinct national social, regulatory, and tax policies. National strategies for growth, development, and employment sufficed.

The economic community that emerged in the 1950s had several fundamental attributes that proved important in the reignition of the European project in the mid-1980s. First, the initial effort was the product of intergovernmental bargains. Second, there was the partial creation of an internal market that is, a reduction but not an elimination of the barriers to internal exchange. The success of this initiative was suggested by the substantial increase in intra-European trade. Third, there was continued toleration of national economic intervention; in fact, in the case of France, such government intervention in domestic economy was an element of the construction. There was an acceptance of national strategies for development and political management. Fourth, the European projects were in fact quite limited, restricted for the most part to managing retrenchment in declining industries and easing dislocations in the rural sector (and consequently managing the politics of agriculture) through the Common Agricultural Policy. There were several significant exceptions, including the European Monetary System that emerged as a Franco-German deal to cope with exchange rate fluctuations that might threaten trade relations; however, the basic principle of national initiative persisted. Fifth, trade remained the crucial link between countries. Joint ventures and other forms of foreign direct investment to penetrate markets continued to be limited. Sixth, American multinationals were accepted, if not welcomed, in each country.

When the global context changed, the European bargains had to be adjusted for new realities. Wallace and Wessels have argued that "even if neither the EC nor EFTA had been invented long before, by the mid-eighties some form of intra-European management would have had to be found to oversee the necessary economic and industrial adjustments."[10]

The structural changes we have been depicting did not "cause" responses. Structural changes pose challenges and opportunities. They present choices to decision makers. Three broad options, individually or in combination, were open to the countries of the EC in the early 1980s. First, each nation could seek its own accommodation through purely national strategies; but for reasons we explore in this chapter, going it alone appeared increasingly unpalatable. Second, Europe could adjust to Japanese power and increase its ties to Japan. It is important to note, though, that Japan cannot provide a solution to the broad range of European problems such as competitiveness, employment, and monetary stability. Britain has cast its manufacturing lot partly with Japanese transplants, but this is not a strategy for the entire continent. Certainly, in specific areas such as electronics technology there are possible industry deals. But that is not the foundation for a broad strategic alliance. The Japanese option also had other significant counts against it: (1) there were no common security interests with Japan to undergird the sorts of relations Europe has had with the United States, and (2) Japan has so far been unwilling to exercise a vigorous leadership role in the international system. The third option was that Europe could attempt to restructure its own position to act more coherently in a changing world. The international changes did not produce 1992; they provoked a rethinking. The 1992 project emerged because the domestic context was propitious and policy entrepreneurs fashioned an elite coalition in favor of it.

POLITICAL ENTREPRENEURSHIP AND THE NEW EUROPEAN BARGAIN

Structural change was a trigger for the original European project; it was a trigger for its renewal. Other factors were equally necessary and, in combination, sufficient for the fundamental bargains to be remade and new, subsidiary bargains to be engaged.

The surprisingly sudden movement by governments and companies toward a joint response does not have a clear and simple explanation. Uncertainty abounds. In a situation so open, so undefined, political science must rediscover the art of politics. The 1992 movement cannot be understood as the unique, logical response to a situation in which actors and groups found themselves, and cannot, therefore, be understood through such formal tools as theories of games or collective action. Neither the payoff from nor preferences for any strategy were—or are yet—clear.[11] European choices have been contingent on leadership, perception, and timing; they ought to be examined as an instance of elites constructing coalitions and institutions in support of new objectives.[12]

This renewal is not a story of mass movements, of pressure groups, or of legislatures. In the 1950s, the European project became a matter of party and group politics. Now, in the 1980s and 1990s, the EC institutions are not the object of debate but are political actors. Indeed, the EC Commission exercised leadership for the renewal in proposing technical measures for the internal market that grabbed the attention of business and government elites but were (in the initial stages at least) of little interest to mass politics. The governments and business elites had already been challenged by the international changes in ways that the parties had not been.[13]

Any explanation of the choice of Europe and its evolution must focus on the actors. The actors—the people who confronted the changes in the international environment and initiated the 1992 process—consisted of the leadership in the institutions of the European Community, in segments of the executive branch of the national governments, and in the business community (principally the heads of the largest companies).[14] Each of these actors was indispensable, and each was involved with the actions of the others. The Commission itself is an entrenched, self-interested advocate of further integration, so its position is no surprise. The multinationals are faced with sharply changed market conditions, and their concerns and reactions are not unexpected. The initiatives came from the EC (originally from the European Parliament and its Draft Treaty on European Union), but they caught hold because the nature of domestic political context had shifted. Tracing the negotiating history of the Single European Act inevitably leads one to focus on the governmental bargains. But the bargaining must be situated in an analysis of what led governments to rethink their conceptions of interests, to redefine their strategies for old goals, and to identify new goals.[15] Casting the analytic net more broadly draws in the Commission and the business elites, as well as the shifting political context of Europe. The interconnections and interactions among the several actors will almost certainly defy an effort to assign primacy, weight, or relative influence.[16]

National governments are at the center of Community politics, and some national governments—particularly the French—have begun to approach old problems in new ways and to make choices that often are unexpected. The question is why national government policies and perspectives have altered. Why, in the decade between the mid-1970s and the mid-1980s, did European governments become open to European-level, market-oriented solutions? The answer has two parts: the failure of national strategies for economic growth and the transformation of the left in European politics.

First, the traditional models of growth and economic management broke down. The old political strategies for the economy seemed to have run out of steam. After the growth of the 1960s, the world economy entered a period of stagflation in the 1970s. As extensive industrialization reached its limits, the existing formulas for national economic development and the political bargains underpinning them had to be revised. Social critics and analysts, in fact, defined the crisis as the failure of established formulas to provide even plausible guides for action.[17] The dynamics of growth and trade changed.[18]

Growth had been based on the shift of resources out of agriculture into industry; industrial development had been based on borrowing from abroad the most advanced technologies that could be obtained and absorbed. Suddenly, many old industrial sectors had to be closed, as was the case with shipbuilding. Others had to be transformed and reorganized: factories continuously upgraded, new machines designed and introduced, and work reorganized. Assessments that eventually emerged acknowledged that the old corporate strategies based on mass production were being forced to give way to strategies of flexibility and adaptability.[19] Despite rising unemployment, the steady pace of improvement in productivity—coupled with the maintenance and sometimes reestablishment of a strong position in production equipment for vital sectors—suggested that Europe's often distinctive and innovative approaches to production were working. However, that was to come only toward the end of the 1970s. In short, national executive and administrative elites found themselves facing new economic problems without adequate models for addressing them.

The 1970s were therefore the era of Europessimism. Europe seemed unable to adjust to the changed circumstances of international growth and competition after the oil shock. At first, the advanced countries stumbled, but then the

United States and Japan seemed to pick themselves back up and to proceed. Japan's growth, which originally had been sustained by expansion within domestic markets, was bolstered by the competitive export orientation of major firms in consumer durables. New approaches to manufacturing created substantial advantages.[20] In the United States, flexibility of the labor market—meaning the ability to fire workers and reduce real wages—seemed to ensure jobs, albeit in services and often at lower wages, despite a deteriorating industrial position in global markets. Japan experienced productivity growth; the United States created jobs. Europe seemed to be doing neither, and feared being left behind by the U.S.–Japanese competition in high technology.

For Europe the critical domestic political issue was jobs, and the problem was said to be labor market rigidity. In some sense that was true, but the rigidities did not lie exclusively or even primarily with worker attitudes. They were embedded in government policy and industrial practice. In most of Western Europe, the basic postwar political bargain involved governmental responsibility for full employment and a welfare net. Consequently, many European companies had neither the flexibility to fire workers or reduce wages of their American counterparts, nor—broadly across Europe—the flexibility Japan displayed in redeploying its labor force.[21] As unemployment rose, the old growth model built on a political settlement in each country was challenged—initially from the left by strategies of nationalization with state investment, and then from the right by strategies of deregulation with privatization. The political basis, in attitude and party coalition, for a more market-oriented approach was being put in place.

For a decade, beginning with the oil shocks, the external environment for Europe was unstable but its basic structure was unchanged. Although the United States was unwilling or unable to ensure a system of fixed exchange rates, it remained the center of the financial system even as it changed the rules. The EMS was an effort to create a zone of currency stability so that the expansion of trade inside Europe could continue. In the 1960s and 1970s, a long debate on technology gaps and the radical extension of American multinational power had not provoked joint European responses. During the 1970s, the mandate for the European Community was not altered, but stretched to preserve its original objectives in the original context. The international economic turbulence and fears of a relative decline in competitive position of the 1970s did not provoke a full-blown European response. The extent of the shifts in relative economic power was not yet apparent. National strategies in many arenas had not yet failed, or, at least, were not yet perceived as having failed. In other arenas, the challenges could be dealt with as accommodations within the realm of domestic politics.

The questions remain: Why did national policy change? Why did the perceptions of choice evolve, the range of options shift? Policy failure must be interpreted and can be assigned many meanings. National perceptions of position are filtered through parties and bureaucracies, shaped and flavored by factions, interests, and lobbies. For instance, in 1983, the French Socialist Party was divided between those led by Laurent Fabius, who concluded that pressure on the franc was a reason to reverse policy direction and to stay within the European Community, and those such as Chevenement, who felt the proper choice was to withdraw from the EMS, even if that resulted in an effective weakening of the Community. The choice to stay in the EMS was by no means a foregone matter.[22] The French response to the currency crisis was a political choice made in the end by the president.

Thus, the second aspect of the changed domestic political context was the shift in government coalitions in a number of EC member states. Certainly, the weakening of the left in some countries and a shift from the communist to the market-socialist left in others helped to make possible a debate about market solutions (including unified European markets) to Europe's dilemma. In Latin Europe, the communist parties weakened as the era of Eurocommunism waned. Spain saw the triumph of Gonzalez's socialists, and their unexpected emergence as advocates of market-led development and entry into the Common Market. Italy experienced a weakening of the position of the communists in the complex mosaic of party positioning. In France, Mitterrand's victory displaced the communists from their primacy on the left. The first two years of the French socialist government proved crucial in turning France away from the quest for economic autonomy. After 1983, Mitterrand embraced a more market-oriented approach and became a vigorous advocate of increased European cooperation. This had the unexpected consequence of engendering independence for the state-named managers of nationalized companies. When the conservative government of Jacques Chirac adopted deregulation as a central policy approach, a second blow was dealt to the authority of the French state in industry. In Britain and Germany, the Labour and Social Democratic parties lost power as well as influence on the national debate.

Throughout the Community, the corporatist temptation waned; that is, management of the macroeconomy through direct negotiations among social groups and the government no longer seemed to work. In many union and left circles, an understanding grew that adaptation of traditional structures would be required. (As the 1992 movement progressed, unions in most countries became wary that the European "competitive imperative" might be used to justify policies that would restrict their influence and unwind their positions and gains.[23] As a counterpoint on the right, Thatcher began to fear a bureaucratized and socialized Europe.)

In an era when deregulation—the freeing of the market—became the fad, it made intuitive sense to extend the European internal market as a response to all ailments. Moreover, some governments, or some elites within nations, felt they could achieve purely domestic goals by using European agreements to limit or constrain national policy choices. The EMS is not only a means of stabilizing exchange rates to facilitate trade, but also a constraint on domestic politics that pushes toward more restrictive macroeconomic policies than would have otherwise been adopted. There is little doubt that the course of France's social experiment in 1981 would have been different if it had not been a member of the EMS, which required a country to withdraw formally from commitments if it wanted to pursue independent expansionary policies. Some Italians used the threat of European competitive pressures to initiate reform in the administration. As one Italian commentator put it, "Europe for us will be providential. . . . The French and Germans love 1992 because each thinks it can be the key country in Europe. The most we can hope for is that 1992 straightens us out."[24]

In any case, Europe witnessed the creation of like-minded elites and alliances that at first blush appear improbable—such as Mitterrand and Thatcher committed to some sort of European strategy. These elites are similar in political function (though not in political basis) to the cross-national Christian Democratic alliance that emerged after World War II in Germany, France, and elsewhere in support of the original Community. European-level, market-oriented solutions became acceptable once again.

This was the domestic political soil into which the initiatives emerging from the European Parliament, and the Commission, fell. Traditional models of economic growth appeared to have played themselves out, and the left had been transformed in such a way that socialist parties began to seek market-oriented solutions to economic ills. In this setting, the European Community provided more than the mechanisms of intergovernmental negotiation. The Eurocracy was a standing constituency and permanent advocate of European solutions and greater unity. Proposals from the Parliament, taken up by the Commission (the EC's administrative arm) and modified by the Council, transformed this new orientation into policy and, more importantly, into a policy perspective and direction. The Commission perceived the international structural changes and the failure of existing national strategies, and seized the initiative.

To understand how European initiatives led governments to step beyond failed national policy, let us examine the case of telematics, the economically crucial sector combining microelectronics, computers, and telecommunications. By 1980, European policymakers were beginning to realize that the national champion strategies of the past decade or so had failed to reverse the steady international decline of European telematics industries. Throughout the 1970s, each national government in Europe had sought to build up domestic firms capable of competing with the American giants. The state encouraged or engineered mergers, and provided research and development grants; state procurement heavily favored the domestic firms. But by 1980, none of these approaches had paid off. Europe's champions were losing market shares both in Europe and worldwide, and most of them were operating in the red. Even Europe's traditional electronics stronghold, telecommunications equipment, was showing signs of weakness: The telecommunications trade surplus was declining annually, whereas U.S. and Japanese imports were accounting for ever-larger shares of the most technologically advanced market segments.[25]

In telematics, European collaboration emerged when the Commission, under the leadership of Commissioner Etienne Davignon, struck an alliance with the twelve major electronics companies in the EC. The twelve firms designed the European Strategic Programme for Research and Development in Information Technology (ESPRIT) and then sold it to their governments. The RACE program (Research for Advanced Communications-technologies in Europe) emerged via a similar process.[26] In short, the Community's high-technology programs of the early 1980s took shape in a setting in which previous national policies had been discredited; the Commission advanced concrete proposals, and industry lent essential support. In a sense, the telematics cases prefigured the 1992 movement and displayed the same configuration of political actors: the Commission, certain political leaders and specific agencies within the national governments, and senior business leaders.

The Commission took the initiative again with the publication of its "White Paper" in June 1985. This initiative was a response to the stagnation of the Community enterprise as a result of, among other things, the budget stalemates. When Jacques Delors took office as President of the European Commission in 1985, he consciously sought an undertaking, a vision, that would reignite the European idea. He seized upon internal market proposals that had circulated between the European Council and the Commission in the early 1980s. The notion of a single market by 1992 caught the imagination because the need for a broader Europe was perceived outside the Commission. Wallace and Wessels suggest that if the EEC and the European Free Trade Association (EFTA) had not existed by the late 1980s, they would have had to be invented.[27] Or, as was the case, reinvented.

The White Paper set out a program and a timetable for the completion of the fully unified internal market.[28] The now famous set of 279 legislative proposals to eliminate obstacles to the free functioning of the market, as well as the analyses that led up to and followed it, expressed a clear perception of Europe's position.[29] European decline or the necessities of international competitiveness (choose your own phrasing) require—in this view—the creation of a continental market.

The White Paper's program had the political advantage of setting forth concrete steps and a deadline. The difficult political questions could be obscured by focusing on the mission and by reducing issues to a series of ostensibly technical steps. Advocates of market unification could emphasize highly specific, concrete, seemingly innocuous, and long overdue objectives rather than their consequences.[30] In a sense, the tactic was to move above and below the level of controversy. The broad mission was agreed to; the technical steps were unobjectionable. Of course, there was a middle ground where questions about the precise form of Europe, the allocation of gain and pain in the process, became evident. A small, seemingly innocuous change in, say, health and safety rules may eventually prove to be the shelter behind which a national firm is hiding from European and global competitors. Here we find the disputes about outcomes, in terms of both market results and social values. Obscuring issues and interests was crucial in developing Europe the first time, one might note, and has been instrumental once again.

Implementation of the White Paper required a separate initiative: the limitation, expressed in the Single European Act of 1985, of national vetoes over Community decisions. At its core, the Community has always been a mechanism for governments to bargain. It has certainly not been a nation-state, and only a loose kind of federalism. Real decisions have been made in the Council by representatives of national governments. The Commissioners (the department heads) are drawn from a pool nominated by the governments. Broader representative institutions have played only a fictive (or, more generously, a secondary) role. Moreover, decisions taken by the Council on major issues had to be unanimous, providing each government a veto. For this reason, it has been difficult to extend the Community's authority; changing the rules of finance, or proceeding with the creation of a unified market and changing the rules of business in Europe, has been painfully difficult. The most reluctant state prevailed. Furthermore, domestic groups could block Community action by persuading their government to exercise the veto.

Many see the Single European Act as the most important amendment to the Treaty of Rome since the latter was adopted in 1957.[31] This act replaced the Treaty of Rome's original requirement that decisions be taken by unanimity with a qualified majority requirement for certain measures that have as their object the establishment and functioning of the internal market. The national veto still exists in other domains, but most of the 279 directives for 1992 can be adopted by qualified majority. As a result, disgruntled domestic interest groups have lost a source of leverage on their governments; the national veto no longer carries the clout it once did. Perhaps equally important, the Single European Act embodies a new strategy toward national standards that were an obstacle to trade within the Community. Previously, the EEC pinned its hopes on "harmonization," a process by which national governments would adopt "Euronorms" prepared by the Commission. The Single European Act instead adopts the principle affirmed in the famous Cassis de Dijon case. That principle holds that standards (for foodstuffs, safety, health, and so on) that prevail in one country must be recognized by the others as sufficient.

The third major actor in the story (the first two being the governments and the Commission) is the leadership of European multinational corporations. In some ways, they have experienced the consequences of the international economic changes most directly. They have acted both politically and in the market. The White Paper and the Single European Act gave the appearance of changes in the EC market that were irreversible and politically unstoppable. Businesses have acted on that belief. Politically, they have taken up the banner of 1992, collaborating with the Commission and exerting substantial influence on their governments. The significance of the role of business, and of its collaboration with the Commission, must not be underestimated. European business and the Commission may be said to have together bypassed national governmental processes and shaped an agenda that compelled attention and action.

Substantial support for the initiatives from Brussels has come from the Roundtable of European Industrialists, an association of some of Europe's largest and most influential corporations, including Philips, Siemens, Olivetti,

GEC, Daimler Benz, Volvo, Fiat, Bosch, ABB, and Ciba-Geigy. Indeed, when Jacques Delors, prior to assuming the presidency of the Commission in 1985, began campaigning for the unified internal market, European industrialists were ahead of him. Wisse Dekker of Philips and Jacques Solvay of Belgium's Solvay Chemical Company in particular were vigorously arguing for unification of the EC's fragmented markets.[32] In the early 1980s, a booklet published by Philips proposed urgent action on the internal market. "There is really no choice," it argued, "and the only option left for the Community is to achieve the goals laid down in the Treaty of Rome. Only in this way can industry compete globally, by exploiting economies of scale, for what will then be the biggest home market in the world today: The European Community home market."[33]

It is difficult, though, to judge whether the business community influenced Europe to pursue an internal market strategy or was itself constituted as a political interest group by Community action. Business began to organize in 1983 when the Roundtable of European Industrialists formed under the chairmanship of Pehr Gyllenhammer, of Volvo. Many of the original business discussions included senior Community bureaucrats; in fact, Etienne Davignon reportedly recruited most of the members of the original group. The executives comprising the Roundtable (numbering twenty-nine by mid-1987) were among the most powerful industrialists in Europe, including the non-EEC countries. The group initially published three reports: one on the need for development of a Europe-wide traffic infrastructure, one containing proposals for Europe's unemployment crisis, and one, *Changing Scales,* describing the economies of scale that would benefit European businesses in a truly unified market.[34]

The European Roundtable became a powerful lobby vis-à-vis the national governments.[35] One member of the Delors cabinet in Brussels has declared, "These men are very powerful and dynamic . . . when necessary they can ring up their own prime ministers and make their case."[36] Delors himself has said, "We count on business leaders for support."[37] Local and regional chambers of commerce have helped to establish about fifty European Information Centers to handle queries and publicize 1992.[38] In short, the 1992 process is repeating the pattern established by ESPRIT: major businesses allied with the Commission to lobby governments that were already seeking to adapt to the changed international structure.

At the same time that the business community has supported the political initiatives behind the 1992 movement, it has been acting in the marketplace. A series of business deals, ventures, and mergers form a critical part of the 1992 movement. Even if nothing more happens in the 1992 process, the face of business competition in Europe is changing. The structure of competition is shifting.

There has been a huge surge in joint ventures, share swapping, and mergers in Europe. Many are justified on the grounds of preparing for a unified market—some for reasons of production and marketing strategies, and some as a means of defense against takeovers.[39] But much of the movement is a response to business problems that would exist in any case. Still, the process has taken on a life of its own. The mergers provoke responses in the form of other business alliances; the responding alliances appear the more urgent because of the political rhetoric. As the Europeans join together, American and Japanese firms scurry to put their own alliances into place and to rearrange their activities.

A few examples will illustrate the nature of business choices and strategies that both respond to and reinforce the 1992 drive. An Italian entrepreneur, de Benedetti, tried to take over the Société Générale de Belgique. The Société Générale, though, is not simply another bank; it is an institution that played a vital role in the 19th-century development of Belgium. The bank has its fingers in a substantial chunk of Belgian industry. The attempted foreign takeover was blocked by an alliance led by the French Bank of Suez with encouragement from the French government. The Société Générale de Belgique became a European development bank, controlled at once by Belgian, French, Italian, and British interests.

A second instance is the recent takeover of Plessey, a British telecommunications and electronics firm, by GEC of Britain and Siemens of Germany. The cross-national nature of the move is itself significant. The acquisition of Plessey by GEC alone had been blocked by the British monopolies commission in an earlier bid, but in the European context, the takeover proved acceptable. A third case is the merging of the semiconductor interests of the French firm Thomson with the Italian firm SGS. What is surprising was that it happened at the same time that Matra was backing out of the semiconductor business. In our view, French electronics is now entirely dependent on foreign sources for the final development and production of microelectronics, though French industry's ties within Europe have been strengthened. As a business deal it makes eminent sense, and was a virtual necessity. That, however, has not always been decisive in French electronics policy. Whether the government simply acquiesced in management's choice or actively promoted the deal, the merger represents a real change in the government's attitude and policy: A European partnership is now acceptable in France. Finally, Asea of Sweden and Brown Boveri of Switzerland, two giants in the electrical-generation business, have merged to form ABB. Swedes, and a Swedish chairman, are now managing the business from Zurich and implanting themselves in the United States through the purchase of elements of Westinghouse. In a direct response, Britain's GEC and Alsthom of the French CGE group have merged their power-generation businesses.

Of course, one must not exaggerate. Deals linking European firms were roughly equal in number and value to those joining European to American enterprises. And the volume of mergers and acquisitions within the United States far outstripped that involving European firms. Still, the boom in intra-European mergers and alliances is a new development. The number of intra-EC mergers, joint ventures, and minority acquisitions involving the 1,000 largest industrial companies in the Community jumped from about 50 in 1983–84 to 180 in 1987–88.[40]

These deals clearly represent decisions by major companies to join together on a European scale in order to position themselves for global competition. In many sectors, as Stephen Cohen points out, 1992 may consist fundamentally of these business alliances and mergers; that is, even if the process is limited to these alliances, big business in Europe will have been transformed. The business deals also represent a change in governmental attitudes to accept and encourage that process. Whereas the pace of European mergers was accelerating by the mid-1980s, it became a rush in 1987. Perhaps not by accident, that was the year in which the political initiatives for a unified market became fully believable.

Open Trade and a Regional Economy?

The 1992 process may have been sparked by a change in the global system; its outcome is certain to shape both the structure and the regimes of the international system in the coming decades. The final shape of the deal that is emerging in Europe is not yet evident, but the new bargain—resting on the White Paper, the Single European Act, mergers and acquisitions, and now the Maastricht Treaty—does represent a sharp change from the first integration project. Recall our specification of the original construction and note that each of its elements is changing. The internal market is being extended. National interventions are being limited and openly constrained by European institutions. The Commission is taking the political initiative in diverse areas from tax and regulatory policy through technology development. Direct foreign investment and joint ventures, not just trade, are forming the links among countries. The real issue here is whether the European Communities will pursue their objectives of a greater role in the world through projects and policies that go beyond the unified market. The vision of 1992 implies the need for future bargains—both internal and external—that are outside the formal 1992 process and its 279 technical measures. They consist of a diverse set of issues whose resolution will determine not only how complete the market unification is, but, more broadly, its character.

Some issues are purely internal. For example, the emergence of a single market pushed forward redistributive and social questions of substantial importance. The structural adjustment funds are the payoff from the richer north to the poorer south for its sustained participation in the project. An essential part of the bargain is a massive increase in funds for the poorer regions, mostly in the south. The regional funds roughly doubled in volume from 1988 to 1993. The estimates are that these funds will amount to as much as 5 percent of Greece's GNP and 1.5 percent of Spain's GNP.[41]

Then there is the related problem of social Europe. Some fear that the very system of social protection and welfare that has become the cornerstone of political bargains in many of the European nations will be put in question. Even the business community is aware and concerned. As one conservative European business leader remarked, "One of the risks is in the social arena. If, in effect, we give in to the temptation to harmonize work conditions throughout Europe, all that we have succeeded in developing and modernizing will be put into question."[42] The EEC, as we noted, has generally set rules about *business* behavior. Except in agriculture, it has rarely directly affected outcomes or directly altered the welfare of specific groups in the society. A genuine internal market without restriction or subsidy will make social policy and tax rules appear, whether they do or not, to affect industry outcomes. Domestic social policy will become an issue of internal market negotiation. The shift of critical social issues to Brussels raised the question for some analysts of whether Europe 1992 was really a device for rightwing liberals and businesspeople to break the corporatist deals with labor that had been built in Europe over generations.[43] The tentative resolution at Maastricht, a bargain of eleven to proceed leaving Britain on the sidelines, is ambiguous at best and perhaps not legal in any case. They see a pluralist lobby emerging in Brussels that will reduce the power of labor. Whether that fear is founded, the question remains of the balance between Community and national decision making. If Brussels begins to make substantial numbers of decisions with visible implications for people's lives, what mechanisms of political control will be established? The list of vital internal issues is long.

As the internal deals were being recast in Europe, those outside watched with concern. They wondered what the overall shape of the deal would be and how it would affect them. They asked whether Europe would move toward a unified and coherent position in the global economy. They speculated on what a unified European position would be; and on what sort of international economic order a new Europe would pursue. The concerns were real for, as a protagonist, Europe would be a major player.

The advanced countries will almost certainly become three trading regions organized around the United States, Japan, and the European Community. Europe as a relatively self-contained economic unit already exists. For nearly twenty years, Western Europe as a whole has represented roughly one-quarter of global gross national product (the European Community, over 20 percent), the United States roughly the same, and Japan about half the share of the United States or Europe.[44] As Chapter 1 of this book shows, our understanding of the global trading system is radically different if we consider Europe as a single unit rather than the sum of its constituent elements. But should we do so? There are two competing images of Europe. One image presents Europe as a set of small and medium-sized countries that have opened themselves to the global economy and must adjust to it. If taken nation by nation (with trade within the region included), Western Europe in 1986 had 44.6 percent of global exports (up from 42.4 percent in 1967) and roughly 42.6 percent of global imports (down from 44.1 percent in 1967). France (26.8 percent) and Germany (32.1 percent) define one end while Belgium (87 percent) and the Netherlands (58.9 percent) represent the other, with Britain (41.2 percent) falling in between.[45] Thus, foreign trade (imports and exports) constitutes a large percentage of domestic product for each of the European countries.

The other image shows the European nations moving over the past thirty years from interlinked national economies to a regional economy. Europe's per capita imports from outside the region are, according to some sources, even lower than Japan's.[46] If we exclude intra-European trade (as we would trade between California and Michigan from American trade statistics), the picture changes. Lafay and Herzog indicated that Europe represents 13.8 percent of global exports in 1986 (down from 15.3 percent in 1967) and 11 percent of global imports (down from 17.0 percent in 1967.[47] In this second view, the countries of Europe are, together, no longer passive takers in the system; they are able to shape their international environment.

The new Europe, as Lafay and Herzog emphasize, lies between these two images. It consists of one tight bloc, the Community, and a looser confederation, the European Free Trade Association. Increasingly, the EFTA is adjusting to recent EEC initiatives, the most significant manifestation being the EC-EFTA accord on a European Economic Area. Sweden has announced that it will apply for EC membership, and Norway and Finland may follow. What the mechanism may be—from status quo through full membership—does not much matter for the central argument here. The political boundaries are beginning to correspond to the existing pattern of economic and trade policies. This shift does not mean that Europe will become a fortress, but, in itself, the creation of a politically unified trade region, capable of coherent action, is significant. Europe will be concerned with itself. It does mean that Europe will now consciously develop joint policies to benefit the Community, policies in which internal considerations are primary.

Europe's choices will not be made in isolation. Europe's external policy will depend as much on choices made by its trading and financial partners as on its own predilections and internal politics. The structural shifts that compel European adjustments are also changing American and Japanese trade policies, changes that in their turn will alter Europe's choices.

The continuing process of a shifting economic order, American political choices and Japanese exports and investments, will set the context for Europe's decisions. American trade with Europe is quite balanced and, consequently, individual trade tensions can probably be managed within mutual commitments to open trade. However, the politics of American trade do suggest change, and real choices, for Europe. American trade legistration now represents a fundamental challenge to the intellectual premises on which policy has rested even while the implementation preserves, for the moment, the substance of U.S. policy. Since World War II, the United States has sought a world of open flows of goods and finance, but not of labor.[48] The principles of multilateral negotiation, extending to all countries concessions made to one, encouraged a focus on rules that would promote the general welfare. "Reciprocal" concessions were required from all parties; reciprocity implied rules for the game, not specifications of the score. Put differently, the United States sought to promote market processes and to avoid bargains about outcomes; the reciprocity was "generalized" reciprocity.[49] The new intellectual frame includes the notion of specific reciprocity: that trade is fair when the outcomes are balanced, and that outcomes—not process—are the core of discussion.

Specific reciprocity emerges in two contexts. In the broadest context, it tries to place the burden for the American trade deficit on the countries with trade surpluses, on the premise that limits to American exports and subsidy of foreign product are the core of the problem. Of course, those national policies to which the United States now objects were in place during the thirty-plus years in which this country dominated trade. They did not produce a defeat then and are hardly the cause of it now. A somewhat narrower context is in telecommunications, the notion being that closed markets in Japan alter the dynamics of competition in advanced technology. The American policy debate thus pushes a new kind of reciprocity (and a debate about outcomes) onto the negotiating table.

Will Pressure from Japan Shape the External Bargains?[50]

By contrast, the competitive pressure from Japan will continue to determine many of the trade choices Europe must make, as Michael Borrus and John Zysman have argued in Gregory F. Treverton's, *The Shape of the New Europe*. Those choices will, almost inevitably, determine whether, or in what way, Europe becomes an economic region turned in on itself—a fortress. Europeans, focused on their own commitments to maintain an open trade system, sometimes ignore the fact that intense competitive pressure from Japan and the rest of Asia may produce an outcome that they themselves do not want. And Americans, on the other hand, often are so focused on the European response to these pressures that they underestimate Europe's commitment to open trade.

A long list can be made of Euro–Japanese trade conflicts, each involving a different type of problem that could affect the character of the European project. When Willy de Clerq was external relations commissioner, his call for reciprocity in financial relations set off a great furor. The policy was aimed at opening access to Japanese financial markets for European firms, or at least limiting a flood of powerful Japanese banks into Europe, but it hit most directly the interests of American financial institutions. Charges of dumping brought by the Community against Japanese firms producing photocopiers in the United States forced the questions both of what entry strategies Europe would accept from Japanese firms and what indeed constituted a Japanese product. Remedies that specified local European content also hit U.S. companies by forcing the Japanese copier companies to design out U.S.-manufactured components. The current debate on Japanese auto quotas in Europe may determine how extensive the trade protectionism that Europe adopts to limit Japanese market penetration will be, and the use of industrial standards to maintain markets becomes evident in the discussion of High-Definition Television (HDTV). One European senior executive, when asked how many HDTV standards there would be, remarked that there would be as many standards as were required to keep the Japanese out of Europe. However, a detailed look at any of these questions—reciprocity, dumping, quotas, standards—would not identify the central dynamic, which is the enormous competitive challenge from Japan.

Japan's industrial success is built on real production breakthroughs and organizational innovations that present exceptional problems for established European firms. The extraordinary profitability of the Japanese firms and the considerable export surpluses that have resulted from their industrial competitive advantage give Japanese firms and financial institutions considerable financial muscle. That financial strength then gives Japanese firms the leverage to implant themselves in European markets.

Although Japanese firms in a broad range of industries are strong in global markets, it is in volume consumer durables such as automobiles and consumer electronics that unique advantages have been established. The effective introduction of "flexible automation" (or a "lean production system") by many, but not all, Japanese firms has given them absolute advantages of cost, quality, and new-product development.[51] The best Japanese auto companies, for example, don't just produce cars less expensively, or produce cars of higher quality; they produce cars that are less expensive *and* of higher quality. Product cycles are shorter; the time from design idea to showroom floor is quicker. This translates into a capacity to respond to market shifts and introduce technological innovation at a faster pace. The result is not only that volume car companies such as Volkswagen, Fiat, and Renault are vulnerable, but, even the top-end specialists, such as BMW and Mercedes, have few automatic defenses against this surge. The same market pressure exists in consumer electronics from VCRs to compact disc players to radios to fax machines to telephone terminal equipment. Certainly, we do not mean to exaggerate by emphasis. Design and innovation advantages from computer architectures through antilock brakes do provide competitive advantages for European and American firms. In chemicals and machine tools, for example, Japanese firms are strong but not globally dominant. Yet the advantages they already hold mean that established firms in European and American markets lose position, and when market share is lost to imports, jobs are lost as well.

Equally important, as Chapter 1 argued, high-volume consumer durables are increasingly becoming the driving force behind technological innovation in advanced technology. High-volume high technology is a crucial new category. High-volume electronic products are increasingly driving the development of a broad range of underlying component technologies such as semiconductors, optoelectronics, precision mechanical and magneto components and displays, and power supplies. Control of these technologies is critical to all electronics production. In automobiles, similarly, electronic engine management, antilock braking systems, and communications technology make electronic auto component firms sources of profound technological innovation. In our view, for example, two of the most sophisticated electronics firms in Europe—Intermetall and Bosch—produce for the consumer electronics and automobile markets and are not counted among traditional lists of European firms.

The basic question here is not the scale of Japanese exports or the extent of Japanese foreign direct investment (both measures of Japan's competitive advantages), but the character of the European response to this new industrial challenge. First, how much room will Europe provide for Japanese firms? The specific debate in the news is about the length and extent of protection for European automobile producers, but the general issue is whether the pace of Japanese entry will be contained by trade policy. Second, what will be the form, and who will be the agent, of public support for the adjustment of European firms? Precisely because the Commission is trying to uproot state subsidy to national firms, trade protection is all the more probable—despite protestations to the contrary. Third, what Japanese development policies and corporate entry strategies will Europe accept? Here we find the question of dumping restrictions, insistence on local production, and demands for entry to the Japanese market.

There is no single European answer to these questions. There is certainly no common position among the governments, nor, is there a single view within the several departments in the Commission. The range of views among the countries is extreme. The Italians and French have sought in autos and consumer electronics to restrict severely Japanese access to the European market. The precise instruments and policy disputes are adapted to that general purpose. The British, by contrast, see Japanese implants as a mechanism to rebuild Britain's manufacturing base. The British see Japanese FDI not just as investment but as an opportunity to learn new notions of how labor and management can operate and how to reorganize production. Other national views fall between on this spectrum, varying degrees of toleration of open competition depending on the basic competitive position of the nation's industries. Thus, German opposition to the auto quotas reflected the small but highly profitable position of Mercedes and BMW in Japan and the confidence of these companies that they can hold their European market. But the position is reported to have shifted shortly after a careful evaluation of the Lexus (Toyota) and Infiniti (Nissan) entries into the luxury market that, along with Honda's NSX entry, provide a real challenge to Europe's leading niche producers. (For those who doubt this, note the October 1990 issue of *Car and Driver* that rates the four leading sedans at any price. The Lexus 400 made the list at No. 3, at roughly one-half to one-quarter the price of the others.) Similarly, there is a broad range of views within the EEC Commission. Consider the notion of a development strategy for European electronics represented by the position of DG XIII, the Directorate General for Technology. That view is substantially challenged by DG IV (Competition), which has gained influence and will continue to seek to wrestle power away from both the rest of the EEC bureaucracy and the member states.

Although for now there may be no single European policy strategy or view, there may be an outcome generated from the initial mix of policies. That initial outcome, we propose, will be that the EC will force foreign direct investment and local production. That is, Japanese firms will be ceded market share more easily when that share comes from production within Europe, rather than from imports. The view rests on four explicit assumptions, each of which is plausible but may be seriously mistaken. First, and centrally, is the notion that if firms are implanted in Europe they can be controlled, influenced, or instructed by national governments and the Commission. On matters of plant location or employment policy this may be the case, but on matters of technology development and diffusion it is much less certain. Second, production in Europe will at least limit employment loss. For many purposes—wages and taxes among them—employment in a Japanese plant in Europe is as good as in a European-owned plant and not all that different from employment in an American-owned plant. However, the real levels of employment may be radically reduced if new, more efficient production organization is introduced and European industries restructure. Third, European producers contend that if Japanese companies operated under European social rules and union environments, they would lose their vaunted production edge. However, Japanese firms operating in Europe and the United States are able to achieve productivity and quality rates well above those of local competitors even though they do not match the performances that they achieve in Japan. Therefore, forcing transplants may not put the Japanese at a disadvantage.

Fourth, it is proposed that if local-content rules and high tariffs on components and subsystems compel the sourcing of parts in Europe, then European producers will hold onto much of the value added in products even if final assembly moves to Japanese hands. Unfortunately, the American experience suggests strongly that local suppliers will have considerable difficulty fitting into Japanese systems and, often, meeting Japanese standards. The result will be that the final product assemblers will pull their Japanese suppliers with them into Europe. This raises the central question of how Japanese entry will affect the supply base and development trajectory of Europe. The industrial supply base consists of the set of producers of components, materials, and production equipment. It represents an infrastructure of skills, know-how, and technology. That infrastructure is the base on which next-generation product and process must be built. It contains the capacity to adjust to shifts in the pattern of market demand. Although many of the necessary skills, components, and systems are available on global markets, the capacity of firms in one particular place to adapt in a timely fashion does seem to turn on the locally available knowledge and skills. Customers are able to compel appropriate access, price, and quality from local suppliers far more directly than they can from remote suppliers. This is a particular problem because producers—as Japanese suppliers—are linked to the industrial families of the major Japanese auto and electronics firms that compete in Europe.

Thus, the question becomes whether the entrance of Japanese suppliers brings to Europe new techniques and technologies that then diffuse more broadly into the fabric of the European economy, or whether those suppliers will control technology access, price, and quality to advantage only their major Japanese customers. The outcome, not deducible from analytic premises, turns on which activities are moved to Europe and how product and technology development in Japan is linked to European activities. The advanced electronics segment of the European supply base is the most vulnerable element because it is simultaneously Europe's weakness and Japan's perceived greatest strategic advantage. The recent announcement by Toshiba of new battery technology for portable computer applications from one of its suppliers, and Toshiba's unwillingness to disclose the supplier or permit access to the technology, are suggestive of the potential leverage at stake. These concerns are explicitly articulated in recent European Commission policy positions agreed to among directorates.

Undoubtedly, Europe's industrial position is strong despite the years of supposed sclerosis and Japanese ascent. The strengths continue to lie in volume production, though the Japanese challenge is now emerging in traditional sectors such as textiles, and production equipment in general. European firms have been very effective at applying advanced technology; they often have been at the forefront in developing products embedding advanced technology. But they have been less effective at maintaining a market position in a range of the advanced electronic products from semiconductors through computers. The visible weakness lies in advanced technology, but when that category is disaggregated, there is one fundamental weakness—advanced electronics. The series of joint research efforts, including Esprit, Race, and Prometheus, are clearly advanced electronics technology programs developed in response to this perceived weakness.

Fujitsu's takeover of ICL underscores the weakness of the European electronics base. In one sense, the sale simply reemphasizes the market weakness of most of Europe's computer firms; apart from Siemens, neither Olivetti, Bull, nor Philips is in a fundamentally more sound position than ICL. It also indicates that due to critical weakness in their internal product portfolios, European computer firms are dangerously close to being nothing more than distributors for Japanese producers. However, the ICL takeover is more than a symbol. Because joint European R&D programs were set up as a response to the European weakness, there has been a resistance to American participation. Some American firms that both produce and engage in R&D in Europe have joined some EC programs. Now, of course, the question becomes whether these institutional rubrics can support the European effort to create a technological response to Japan. The same logic that led Sematech to exclude all foreign companies from its efforts in the United States—the perceived need to keep the Japanese out—is posed here in Europe.

The European technology support programs are being built at the same time that an effort is being made to dismantle programs of state subsidy to domestic producers. The decisions to compel IRI's subsidiary, Finemeccanica, to reorder the books of the sale of Alfa Romeo to Fiat is but one instance in a series of such decisions. National subsidy has often served as an alternative to formal trade restrictions. It is a floating and indirect tariff, but one that avoids formally redefining the character of the trade relations. Will formal bilateral restrictions on trade be more likely in the absence of domestic subsidy? We suspect so.

This leaves one final issue: the Japanese developmental system and the Japanese market for imports. Europe is responding to Japanese entry, as the

United States did, with concerns over the Japanese government's strategy to help create advantage in global markets and the restrictions on exports to Japan. Japanese policy, by contrast, has moved from development strategy language of restriction toward a language of liberalism. The result has been a loosening of Japan's developmental system, but, in our view, not its abandonment.[52] Markets have been formally opened, but primarily where Japan has an established advantage in global markets. In sectors where advantage is yet to be established or where Japanese firms have lost the advantage, a band of protectionism still remains. Barriers remain both in policy and in business practice. Their significance lies in several domains. First, in emerging or transforming sectors: If one powerful national player maintains a closed domestic market, its firms gain substantial competitive leverage. Second, if the United States is not able or willing to absorb the world's exports, will Japan do so? Third, if barriers of access to financial markets are maintained, then Japan gains leverage in global financial and product markets. Japanese financial institutions can then use their role in intermediating the massive Japanese savings to entrench themselves in the financial markets of the United States and Europe. European politicans such as Rocard and Thatcher, divided as they are on other issues, have nonetheless teamed up to argue powerfully that European retaliation is inevitable unless Japan opens its markets and restrains its use of policies and pricing to create advantage. Europe's stance vis-à-vis Japanese entry will, in the end, be influenced by its own success in Japanese markets.

Summary

These American and Japanese developments press on Europe questions that it must resolve in any case—those of reciprocity and outcomes on the one hand, and the choice between free market and developmental policies on the other. The notion of general reciprocity in trade bargains, which focuses on maintaining market processes, gives way to specific reciprocity, which focuses on specific outcomes. Most concretely, there are manufacturing sectors where national restrictions must either be abandoned or generalized to Europe. The real challenge to national producers is not, for the most part, other European or American producers, but Japanese producers. Consequently, in a range of products from automobiles to dot-matrix printers, the question becomes the terms on which the European market will be available to Japan. For instance, at present, antidumping rules are applied to restrict imported printers and photocopiers. In automobiles, an EC–Japan agreement reached in mid-1991, however, leaves cloudy the issue of access to the EC market for Japanese automakers. The understanding "forecasts" a Japanese market share of 16.1 percent by 1999 (up from 9.4 percent in 1989). The two parties apparently agreed that Japan would limit vehicle exports to the EC. But whereas EC negotiators claim that the deal also restricts the output of Japanese "transplants," the Japanese disagree. In any case, the deal looks like a negotiated allocation of market shares. 1992 is accelerating Japanese entry into the European market.

America initially reacted nervously to the 1992 project, fearing a "fortress Europe" that would shut out American products. Events have mostly allayed those fears because the Community has not engaged in the wholesale construction of barriers. Americans have relaxed because protectionist measures in Europe have frequently been directed against Japanese goods, especially electronics and automobiles. Engaged in their own contentious trade disputes with Japan, Americans may be more likely to sympathize with the Europeans than with the Japanese. Still, Europe's newfound assertiveness and cohesion on market matters will make it a formidable player in international trade. Free-trade areas are under discussion in North America and in Asia. In their public rhetoric at least, supporters of the proposed free-trade areas assert the need to respond to Europe 1992.

Europe will be an international protagonist in trade, but in what kind of role? Will the EC promote principles of free trade, or espouse developmental policies to justify protection? Clearly the answers depend both on the positions taken by Europe's major partners and on a complex chain of bargains among EC members. As the Commission cracks down on government subsidies to industry, some states will try to turn to trade restrictions as an alternative. The initial outcome seems to be the creation of incentives for foreign (especially Japanese) producers to locate within the Community. The deadlock in the Uruguay Round of GATT negotiations creates the danger that expanding free trade may take second place in Europe to preserving politically privileged economic programs.

The Second Phase: Redefining Europe

The European project will, for the second time in a decade, be sparked by a change in the international system. Europe is pressed another step toward the role of protagonist on the global stage. The retreat of Soviet power from Eastern Europe forces Europeans to define both who is in the community of democratic, market-oriented nations and what the relations among them will be. With the Cold War now truly over, the direct military threat to Western Europe is reduced. This alters the European security problem and inevitably brings a reconsideration of Europe's fundamental interests. Inevitably, both security and the place of military defense must be reconceived.[53] Whereas the first phase, recasting EC bargains, was a response to essentially economic challenges, the second phase entails a redefinition both of the Community itself and its place in the world.

The revolutions in Eastern Europe, welcome as they were, brought in their wake three major challenges to Community institutions and roles. First, stability in the Central European corridor of Poland, Czechoslovakia, Hungary, and Yugoslavia suddenly became an issue for the EC. Civil war in Yugoslavia turned vague concerns into an immediate crisis, and the EC has led international diplomatic efforts to stop the fighting. The Community had an immediate stake in the economic and political development of Eastern Europe, not least because of the possibility of sudden and massive migrations should reforms fail. And in the long term, those same countries are virtually certain to become EC members. Second, the waning of the Cold War removed one key obstacle to membership applications from the European Free Trade Association. Already fearing exclusion from the EC's internal market project, EFTA countries could actively pursue membership in the EC once Europe was no longer the fulcrum of Cold War balancing. Austria has already applied and Sweden has announced its intention to do so. Finland and Norway may follow. The prospect of new members from EFTA and from the East called into question the viability of Community institutions: If joint decision making was difficult for twelve, it would be impossible for eighteen or twenty. EC institutions would have to be reformed in order to prevent obstructionism and deadlock. "Deepening" the existing Community, both by integrating in new areas (monetary, foreign policy) and by reforming institutions, became the Commission's priority, argued explicitly and repeatedly by Jacques Delors.

The third major challenge driving the redefinition of Europe was unified Germany. German unification raised the same issues that Eastern Europe and EFTA did, only in more immediate form. A consensus rapidly emerged among Germany, its EC partners, and the Commission that the new and enlarged Germany had to be integrated more tightly than ever in an expanded web of Community ties. Thus, the movements for European economic union and political (foreign policy) cooperation accelerated as German unification proceeded. Indeed, monetary union is, above all, a political objective—not an economic necessity. Its importance merits closer analysis.

The fusing of Germany's two halves in 1990 revived questions that had been dormant as long as the Cold War standoff froze Europe's economic, political, and security frontiers. The European Community anchored part of Germany in the West. German unification, in the context of a general Soviet withdrawal from Eastern Europe, gave rise to multiple uncertainties and questions about Germany's ties. Some worried that Germany would weaken its links to the West, either as part of a package negotiated with the Soviets or as a response to opportunities for economic and political leadership to the east. In late 1989, in fact, some observers speculated that Germany would no longer perceive a need for the Community. Another set of concerns focused on the weight that an enlarged Germany would carry in Europe. West Germany had been the region's strongest economic power; a unified Germany might become a hegemon in Europe. It would have ever greater power to shape the regional economic environment.

Monetary union had been on the EC agenda since the 1969 summit at The Hague. It received only indirect mention in the Single European Act, but took on new impetus as the 1992 project forged ahead. With German reunification, the Commission and Community members (especially France) redoubled their effort on behalf of economic and monetary union (EMU). Germany and its partners endorsed EMU as a means to build on the single market project, avert

German preoccupation with its own problems, and avoid German domination of the European economy. EMU would both bind Germany into a Community central banking system (and a single currency) and dilute Bundesbank control of joint monetary policy.

The impetus for EMU is thus clearly political. Indeed, there have always been open disagreements about the economic benefits and details of EMU. Though presented by Jacques Delors as the necessary and natural extension of the single-market project, the economic logic is at best cloudy. Professional economists offer widely divergent assessments of the economic wisdom of monetary union in Europe. Unlike in the cases of free trade and factor mobility, there is no unambiguous answer to the question of whether permanently fixed exchange rates or a single currency is required by a single market.[54] Certainly, the elimination of capital controls in Europe could generate capital flows that might not be manageable within the existing exchange rate system. The assumption is that German monetary policy restricted the options of other countries, unless they resorted to capital controls. Removing capital controls, the logic runs, will amplify pressures on exchange rates. This logic is not unchallenged and the gains from an EMU not unequivocal. Eichengreen, reasoning from the American experience, contends that an EMU would be difficult to sustain unless Community budget transfers are sufficient to offset market shocks within a single economic community.[55] The Community budget is a much smaller percentage of Community GNP than the federal budget is of American GNP. Absent a real economic compulsion or political necessity, formal models suggest that the incentives for countries to participate are very weak.[56] Or, more precisely, the economic incentives are only convergent if small countries have proportionally more influence than larger ones in the decisions of the central bank. Reaching an equilibrium distribution of power is therefore very difficult. The details of this and other models are not crucial. What does matter is that there are both a substantial difference in judgment about the gains and problems of an EMU, and real political obstacles to solving the problem of adherence and governance. EMU will require a significant loss of national policy autonomy to achieve any economic gains. Furthermore, there is no automatic or natural economic process or pressure that will lead to an easy implementation of EMU.

The movement toward EMU must therefore be seen as a political solution to problems other than the single market or financial liberalization. In our view, EMU became an effort by Germany and its neighbors to strengthen and expand the web of Community ties. Furthermore, for Germany's partners, EMU would dilute Germany's preeminence in Europe. If, as many argue, the EMS is really a D-mark zone run by the Bundesbank, then the EMU is an effort by the other countries to have a seat at that monetary table. Thus, German unification mobilized interests that pointed toward an enhanced, not a reduced, role for the European Community.[57] Germany had two overarching motives for fortifying its EC ties in the wake of unification. First, the EC market remains crucial to the German economy, accounting for over half of German exports and imports. Second, German leaders wanted to demonstrate that the enlarged Federal Republic would remain a committed participant in Western (especially European) institutions. Chancellor Kohl repeatedly emphasized that Germany's agenda was unification within the context of European integration, not to seek a hegemonic role in a newly configured Europe.

By the same token, Germany's EC neighbors had powerful motives to ensure that Germany would stay a member of the club. On the economic side, the German market was a crucial part of the common market; the share of total exports flowing to Germany ranged from a low of about 11 percent for Ireland and Spain to highs of 27 and 31 percent for the Netherlands and Greece, respectively.[58] Most of the EC countries had tied their currencies to the Deutschemark in the EMS, benefitting from its antiinflationary powers. On the political side, the more Germany was integrated into common institutions, the more secure its partners would feel. As it became clear that German unification was inevitable, President Mitterrand began pushing for increased EC integration, from monetary union to common security policy. Denmark, so often among the most reluctant of partners, displayed a sudden enthusiasm for strengthening the Community.

In short, both Germany and the other EC states had a powerful interest not only in continued German commitment to the Community but in expanding the number and nature of EC ties. The growing consensus on the need for strengthening the Community found concrete expression in parallel intergovernmental conferences convened in December 1990, one to pursue monetary union and the other to chart the course toward political union. Both projects reveal a European Community trying to redefine itself and its role in the world.

Of course, EMU was already on the EC agenda when German unification became an issue. The Commission had been promoting the idea as the 1992 project pushed ahead. The French endorsed the notion, calling in January 1988 for a European central bank. Thus, the interest in monetary union did grow out of the experience with the EMS and the drive for completing the internal market. Monetary union would complete the former and complement the latter. But the political changes in Eastern Europe—and thus in Germany—gave EMU a decisive new impetus.

As revolution added to revolution in Eastern Europe during 1989, Commission President Jacques Delors repeatedly argued that progress on "deepening" the Community, through monetary union and political reforms, was becoming increasingly urgent. The EC summit at Madrid in June 1989 committed the Community to beginning stage 1 of the Delors plan for monetary integration in July 1990, and called for the preparation of the measures needed to proceed to full monetary union.[59] President Mitterrand and his ministers echoed Delors; in fact, by October, Mitterrand was calling for early discussions on monetary union, asserting that the changes in Eastern Europe meant that the EC "should accelerate its own construction."[60] In early November, to the world's amazement, the Berlin Wall tumbled down. Mitterrand went on a tour of European capitals to drum up support for his proposal for speedy progress on monetary integration.[61] Chancellor Kohl firmly endorsed French proposals to push ahead on political and monetary union. Later that month, Kohl announced his ten-point plan for German federation, dramatically raising the tempo. Meeting at the Strasbourg summit in December, EC leaders endorsed the idea of German unification "in a context of European integration"—a link that Kohl and Foreign Minister Genscher were at pains to reiterate. The Strasbourg summit, not coincidentally, also called for intergovernmental conferences on monetary union and political union to be convened before the end of 1990. In the wake of German developments, the monetary timetable became substantially more concrete and ambitious than the project on political union.

The two halves of Germany agreed early in the spring of 1990 on their own economic and monetary union (GEMU), to take effect on July 1. Some observers, noting that GEMU would entail a fiscal drain on the West and thus exert pressure on the Deutschemark, speculated that Germany would be less interested in monetary union at the EC level. Again, events had the opposite effect. By February 1990, Jacques Delors was calling for institutional reforms (greater powers for the European Parliament, increased majority voting in the Council) and for EC foreign policy. Following the March elections in Germany (electing a Christian Democrat–led government in the east), French Foreign Minister Roland Dumas called for quickening the process of monetary integration and moving on toward political union. In May, Kohl and Mitterrand issued a joint communiqué calling for political union, to begin in 1993. The member states strongly endorsed both monetary and political union; only the British were viscerally opposed to both.

At the Dublin summit in June 1990, the EC heads of state voted to convene the two intergovernmental conferences in December. The conference on monetary union would prepare treaty revisions, to be ratified by the end of 1992. The British found themselves in the anomalous position of taking part in EMU discussions while not participating in the exchange rate mechanism of the EMS. The tension inherent in that anomaly helped bring about British membership in the exchange rate mechanism (in October 1990). At the Rome summit that October, Thatcher was more isolated than ever, as the eleven other countries voted to begin the transitional stage 2 of the Delors plan in January 1993. Thatcher also dissented from the list of topics the other eleven agreed should be discussed at the intergovernmental conference on political cooperation. Thatcher's recalcitrance on EC issues contributed to her downfall less than a month later.[62]

Twin intergovernmental conferences began as scheduled in December 1990. Both sets of negotiations confronted daunting agendas. The conference on political union had to settle questions concerning common foreign and security policies, a joint approach to visas and migration, and an EC-level social charter (on rights and protections for workers). Discussions on monetary union faced formidable challenges, such as designing the transition to a European central bank and a single currency, and the degree of economic "convergence" to be required before countries could take the final step. In the end, both conferences produced treaties that were signed at Maastricht in December 1991.

Regarding political union, the compromises needed to produce a treaty that could be signed meant that the final agreement was less ambitious than its most enthusiastic advocates (the French and Germans) desired. On social policy, since Britain rejected any Community-level policy, the other eleven members signed a separate protocol. Under its terms the eleven can devise joint policies on working conditions (health and safety), worker consultations, and equality of men and women in the workplace. On foreign policy, the Council of Ministers may decide by unanimous vote that a foreign or security policy issue should be the subject of joint action. It may also specify, again by unanimity, which implementing measures can be decided by qualified majority vote. The treaty also gives the Community, for the first time, a role in defense policy. It "shall include all questions related to the security of the European Union, including the eventual framing of a common defense policy, which might in time lead to common defense." The Western European Union (heretofore the European caucus within NATO) will design and carry out defense-related policies when requested by the Community and will serve as a bridge between the EC and NATO.

Though the provisions for common foreign and defense policies are vague, they nevertheless lay the foundation for an independent EC security role in the next century. Most member states seem to believe that the United States will continue to reduce its military presence in Europe and that the Community will therefore be compelled to build a coherent defense structure of its own. Indeed, in early 1992, President Mitterrand and Chancellor Kohl called for a genuine European unit of perhaps 35,000 soldiers. The unit could be used as a rapid response force outside of Europe and could provide the nucleus for a European army.

Of the two Maastricht treaties, that on monetary union is by far the more concrete and dramatic. Eleven states committed themselves to rapid and irreversible movement to a single currency and a common central bank. EMU will, it is asserted, be fully in place by 1999 at the latest, and possibly as early as 1997. During 1996, the European Council (the heads of state or government) may decide by unanimity to move to stage 3 (a single currency and an operational European Central Bank) provided a simple majority of EC states meets the economic criteria for full participation. If the governments do not agree on the final move to EMU by the end of 1997, stage 3 will begin on January 1, 1999, with whichever states meet the economic criteria (whether they comprise a majority or not). The monetary union will be managed by a European Central Bank, which will be independent of national governments and Community authorities and whose priamry goal will be to ensure price stability. National central banks will become independent of their governments before the transition to stage 3, at which time they will become in effect branches of the European System of Central Banks.

The treaty also lays out the economic criteria that states must meet in order to participate fully in the EMU. States moving on to stage 3 will display:

- A rate of inflation in the consumer price index no more than 1.5 percentage points higher than the average of the three states with the best performance in price stability

- Interest rates on long-term goverment bonds no more than 2 percentage points higher than the average of the three countries with the lowest rates

- Central government budget deficits no greater than 3 percent of gross domestic product (GDP)

- Public debt of no more than 60 percent of GDP

- A national currency that has remained within the narrow (2.25 percent) fluctuation margins of the ERM for the previous two years and has not been devalued against any other member state currency over the same period

Member governments have committed themselves to meeting the criteria in time to join the first wave of entrants into full EMU. For some states (Greece, Portugal) that will be virtually impossible and they will make the transition later. Other goverments (especially Italy's) will find it severely challenging to meet the requirements. And once in the EMU, states will have to keep public sector deficits and debt within the guidelines or face penalties provided for in the treaty. Finally, the reluctant British were accommodated with a special protocol allowing them to opt out of stage 3 should Parliament so vote. Denmark received a similar provision, allowing for a referendum on EMU before joining in stage 3. The Danish case differs from the British one, however, in that whereas the British have consistently rejected the notion of a single currency, the Danish government has supported that goal.

Finally, the implications of the Monetary Union in Europe extend beyond the borders of the Community, and the potential gains for the Community depend on the position it will assume on the international scene. In that regard, the European currency (the ECU) will almost certainly, over time, take its place alongside the dollar in trade, official reserves, and financial markets. Certainly EMU will give Europe enormous weight in both monetary negotiations and institutions. Europe voting as a bloc with current quotas would have greater power in the IMF than the United States and Japan combined. Moreover, it would obtain—along with the United States—an effective veto, even if its quotas were reduced by excluding intracommunity trade. Equally important, influence and profit could come to the Europeans by an extensive role of a common currency in world monetary markets.[63] The EMU, and its implications for internal efficiency, cannot be separated from its external role in defining Europe's place in an emerging international financial order in which the dollar may not be the only international money—or international money at all.

Indeed, in finance and trade it does appear that the outcome of phase 1, recasting the bargain, and phase 2, redefining Europe, makes it probable Europe will emerge as a protagonist. But what about the security and political realm?

Europe as Protagonist? Resetting Europe's Place

Europe's adjustment to the retreat of Soviet power has already begun to alter its very definition and purposes. On many issues the alternative to a collective European action will be simple national squabbling, leaving Europe a taker of choices made elsewhere. Some see the war in the Persian Gulf as a harbinger: The United States set the tone and direction of policy, and European governments faced the choice of going along or not. The answer will not be as simple as a single federal executive, but rather a complex pattern in which Europe is likely to find a common position rapidly enough to shape choice on some matters, but not able to do so on others. Again, our view is that the strongest incentive or pressure to generate mechanisms for common European action comes from the change in the global system. The system shift is twofold. First, Europe's security position will not in the future be as closely aligned with the United States. The corridor of former Soviet satellites between Western Europe and Russia reduces the military danger of overrun; the former Union's concern with its own internal difficulties reduces the security threat. But existing security structures in Europe are premised on that threat and on American involvement on the continent defined by a common interest in the face of an aggressive Soviet Union. NATO members agreed to maintain the organization, but the significant fact is that its future existence was up for discussion.

Second, the real dangers to the European nations come in different forms and different places. Instability in the corridor running from Poland through Yugoslavia, to be intensified if conflict erupts among some of the former Soviet republics, will generate problems that only begin with people fleeing from economic dislocation or even from civil war. That instability will be principally a European problem; America has more or less opted out of a role in shaping developments in these regions. The White House has certainly followed the EC lead in dealing with the Yugslavia crisis. Political turbulence in North Africa and the Gulf pose a threat not onlly to oil supplies, but of migration and increased political activism from minority Muslim citizens. The United States has displayed little understanding or sensitivity to these problems, and in any case our common security framework is largely irrelevant to them.

The renewed call by Mitterrand and Kohl in the winter of 1990 for acceleration of discussions of political union, initially voiced in the spring of 1990, came in part as a response to the European inability to shape events in the Gulf, a region that affects Europe in such varied ways. The treaty endorsed at Maastricht builds a framework for a common EC foreign and security policy in the future. The range for joint political maneuver will grow both as Europe senses the extent of its collective resources and as it confronts the consequences of American withdrawal from leadership in areas such as Central Europe. For the United States the dominant question is whether a single European—or at least Community—defense posture is established such that Europe renegotiates its defense alliance with the United States.

Whether or not the Atlantic alliance undergoes formal revision, security ties between the United States and Europe are certain to change. Some of NATO's functions may go the way of old soldiers—they may simply fade away. Europe will redefine its security interests and arrangements. An EC security arm will not emerge fully formed, as a political construction. Rather, the political and defense architecture will be built in response to a series of crises and demands, each posing the need for common action. In the meantime, the intertwining of economic and security matters will produce gradual, piecemeal changes in Europe's security identity.

For instance, high-technology development overlaps military and commercial sectors. At one level, high-tech defense industries are becoming less national and more regional. Cross-border joint projects are already the norm for aircraft, helicopters, and missiles. Networks of alliances link GEC, British Aerospace, MBB, and Aérospatiale. Recent transnational mergers and acquisitions have further integrated the European defense industry. Siemens and GEC purchased Plessey in 1989. GEC and Daimler-Benz (a major defense firm since its purchase of Dornier, MTU, AEG, and MBB) each hold a stake in Matra. As Walker and Gummett have argued, such links will create headaches for policymakers, especially since the major defense firms are also the largest commercial enterprises in aerospace, electronics, and communications. National and EC officials in charge of safeguarding industrial competition will therefore face new challenges in the form of giant crossnational consortia active in both defense and commercial markets. National defense procurement agencies will face the same few broad industry groups, which will therefore have a greater say in which defense systems are to be produced, across Europe. Walker and Gummett make the case that de facto industrial integration could lead to pressures for increased cooperation in procurement decisions, which, in turn, require closer coordinating of defense strategies and force postures.[64]

The 1992 project for the internal market could overlap with procurement issues. Already in place is EC legislation that will open public procurement markets in the previously restricted sectors: transport, telecommunications, water, and electricity. The military will remain one of the largest sectors for government purchases. As liberalization in other state-dominated markets delivers real savings, pressure could mount to provide the same economies in defense procurement.

European collaboration in high-technology R&D could also provide an avenue for increased security cooperation. Community programs such as ESPRIT and RACE have produced solid, though unspectacular, results in microelectronics, computers, and communications technologies. Participants generally regard such programs as well run and worthwhile. The EUREKA program, a collaborative effort in high technology with over $6 billion worth of joint projects, initially had an ambiguous orientation. Though its French proponents stressed that it would be a civilian program, the technological areas outlined for EUREKA exactly paralleled those being pursued in the U.S. Strategic Defense Initiative.[65] To the extent that the boundary between advanced commercial and defense technologies has blurred, such programs may support (or could easily be adapted to support) a European technology base for both civilian and defense applications. Such programs could also provide an organizational model for collaborative defense R&D programs in the future.

In short, Europe faces drastically altered security challenges. The Soviet–American Cold War faceoff no longer dominates European security. The Warsaw Pact dismantled itself, and a buffer of independent (but West-leaning) states emerged between Western Europe and the Soviet Union. At the same time, a European Community with renewed vigor and self-confidence is solidifying its integrative ties and extending them to new areas. As a result, formerly axiomatic definitions of European security are thrown open to question. The new answers will include a more unified, self-consciously European defense presence linked, in the long term, to the European Community. But we should make it clear that for Europe to redefine itself, each country must reassess its own position. That itself could lead to divisions within the Community. It is not impossible that serious divisions could splinter the new unity. The Yugoslav crisis saw such an internal rift. But the fear of disintegration is itself a pressure for further unification. Consequently, this analysis is a judgment of the balance between two possible trajectories.

Conclusion

Europe wrestles with shifts in the international system and with a drastic reconfiguration of its own political landscape. Both the distribution of capabilities and the nature of threats (economic and security) are shifting dramatically. We hypothesize that in the first phase change in the international economic system was necessary (though not sufficient) for the initial revival of the European project in the form of the 1992 project (recasting the EC bargain). We also hypothesize that its intensification and extension to the monetary and political arenas were responses to seismic shifts in the political foundations of postwar Europe, namely, German unification and Soviet withdrawal from Eastern Europe. The sudden enlargement of Germany and the prospect of new applicants from Eastern Europe and EFTA accelerated Community efforts to strengthen EC institutions and pursue monetary integration, a process we called "redefining Europe." The dissolution of the Soviet bloc altered the security context for Europe. In response, Europe is resetting its international position by seeking to define security interests and institutions that will be more autonomously European. The Community is already a major international protagonist in trade, and may become one in monetary affairs.

A full-fledged test of these propositions will require detailed analysis of the perceptions and beliefs of those who participated in launching the 1992 movement and its follow-ons. Certainly the available political histories support these propositions. Structural situations create the context of choice and cast up problems to be resolved, but they do not dictate the decisions and strategies. Changes in an existing distribution of capability and of threat may undermine a particular alliance system, but the alliance system that emerges is a reflection of the perceptions of threat and of the conceptions of how to respond.

In other words, the global setting can be described superficially in neo-realist fashion, but the political processes triggered by changes in the system must be analyzed and explained in other than structural terms. The choices result from political processes and have political explanations. In this case, the process was initially one of bargains among nations and elites within the region. "Europe 1992," the EMU, and a European political presence form a complex web of bargains and accommodations, the key ones being among the national governments, but also involving business elites. Furthermore, the Commision has, at various stages, shaped the agenda, accelerated the negotiations, and guided the bargaining. The outcomes are quite unknowable, dependent on the timing and dynamics of a long series of contingent decisions. But the story, and consequently the analysis, concerns political leadership in creating a common European interest and then constructing a set of bargains that embody that understanding. Many of the choices are simply calculated risks, or perhaps explorations that will be entrenched if they work and refashioned if they don't.

The details may not be possible to foresee, but no one should underestimate Europe's formidable capabilities. Even in an area of perceived weakness such as high technology, Europe's clear problems (weakness in components and computers and the resulting dependence on the United States and Japan) mislead. In militarily relevant high technologies—including aerospace, nuclear energy, advanced materials, transportation, and communications—Europe has major strengths. Airbus has passed McDonnell-Douglas to become the second largest aircraft producer, with a third of the world market. Advanced battlefield communication technology for the U.S. Army comes from Thomson. In high-speed trains, Europe is a leader. In any case, Europe's choices, particularly those that increase the possibility of a coherent Western Europe's emerging as an actor on the global stage, will powerfully influence both the world economic and security systems.

4

From Spin-Off to Spin-On: Redefining the Military's Role in American Technology Development

JAY STOWSKY

Introduction

Soon after the U.S. military's smashing, quick victory in the Persian Gulf war, George Bush went to speak with the people who had built the Patriot missile at Raytheon's defense plant in Massachusetts. "What has taken place here," he said, "is a triumph of American technology; it's a triumph taking place every day, not just here at Raytheon, but in the factories and farms all across America where American workers are pushing forward the bounds of progress, keeping this country strong, firing the engines of economic growth. What happens right here is critical, absolutely critical, to our competitiveness, now and into the next century."

This chapter argues, in essence, that the president gave the right speech in the wrong plant. Far more portentous for American competitiveness than what happens at Raytheon's missile plant may be what has already happened at another Raytheon plant, where the company still produces the Amana radar range, an early version of the microwave oven. A commercial "spin-off" from U.S. military technology, microwave ovens are now a high-volume, technologically sophisticated consumer electronics product. But commercial microwave oven sales for both Japanese and South Korean firms far outstrip Raytheon's sales of today's radar ranges. The growing sophistication of microwave technology continues to increase its use in both military and civilian applications. But it is the employees of East Asian electronics firms, not the employees of Raytheon, who are the most practiced at making it.

From microwave ovens to microchip computers, there now exist two distinct approaches to developing advanced military and civilian technology: Although the United States has traditionally used military projects to generate

114

technological breakthroughs, other countries, most notably Japan, now use commercial markets to accomplish the same ends—faster, with higher standards for product reliability, and at significantly lower cost. In areas where no high-volume civilian markets yet exist—gallium arsenide components, massively parallel computing—technical spin-offs from the U.S. military sector still have time to redound to the competitive benefit of America's civilian firms. But in a world where foreign producers of military-relevant commercial technologies emphasize speed in both product development and technology implementation, time is a luxury the United States may no longer be able to afford.

There are many cases in which the spin-off model has worked well to establish U.S. leadership in both military and civilian applications. In the semiconductor industry, as our first case study will show, military policy certainly helped to clear a path for commercial market penetration. The government required a domestic second source on all Pentagon contracts and provided loan guarantees for constructing new production facilities. Both actions effectively lowered entry barriers and diffused innovative technology among competing firms. Military procurement then provided an extremely effective initial launch market, fueled at premium prices.

Yet even when America's technology edge was at its sharpest, the spin-off strategy sometimes faltered. Our second case study details the ways in which military-specific performance requirements cramped the civilian diffusion of Air Force–sponsored computer control technology for machine tools, encouraging the development of an overspecialized civilian industry that was commercially vulnerable to foreign competitors. In this instance, spin-off proved to be a clumsy mechanism for moving innovative technology from military to civilian markets in a timely and competitive fashion.

By the late 1970s such divergence in performance requirements for military and civilian products had already become a more general phenomenon. More importantly, American efforts to advance the core technologies on which both sets of product applications rested had diverged—unnecessarily—as well. The spin-off approach had created a domestic military-industrial enclave, inhabited by firms that organized themselves for the sole purpose of marketing to the Pentagon. This left them with business strategies and market antennae that were unresponsive to the strains of commercial competition and insensitive to the drift of civilian technological innovation.

Pentagon planners responded with a set of projects designed to extract military-specific applications from state-of-the-art commercial producers. Two (our third and fourth case studies) are examined here—the military's effort to develop very high speed integrated circuits and the Pentagon's Strategic Computing Program sponsored by the Defense Advanced Research Projects Agency (DARPA). In the end, the esoterica of military performance requirements, combined in some cases with wasteful attempts to overcome unnecessary security restrictions, succeeded only in reinforcing the bifurcation that still characterizes the American technology base. New military applications were in fact created, sometimes in a way that genuinely advanced the technology base, but those advances were few in comparison to the rapid-fire achievements emerging simultaneously from the civilian sector—not only at home but, increasingly, abroad. In the end, the military programs did not impede such advances so much as bypass them altogether.[1]

As has already been suggested, it is often the case now that technology diffuses from the civilian sector to the military, rather than the other way around. "Spin-on," an alternative approach to building military and commercial applications from a common technology base, has emerged most fully in Japan, where militarily-relevant subsystem, component, machinery, and materials technologies are rehearsed and refined on high-volume commercial applications. Much of that work occurs in the context of government-orchestrated research projects whose explicit object is the creation of commercial technology applications. Japan's largest defense contractors are much less isolated from commercial practice than are their counterparts in the United States. Mitsubishi Heavy Industries, Japan's top defense contractor, derives just 21.3 percent of its revenues from defense—compared to a typical U.S. prime contractor, such as General Dynamics, whose defense sales account for 68.3 percent of its revenues.

In this new competitive context, spin-off and spin-on must be regarded as different approaches to the problem of organizing and financing the development of advanced technology. As the case studies in this chapter suggest, spin-off might still work under specific circumstances. In areas where no significant civilian competition has yet been established, for example, spin-off could work to diffuse innovative technology from the military to the civilian sector. It would only work, however, because American firms in the civilian sector would have time, in these cases, to appropriate the economic benefits from their own applications.

But now as in the past, spin-off can also be a source of competitive disadvantage. Overreliance on spin-offs hurts civilian economic competitiveness when commercially irrelevant performance requirements are already designed into the technology that diffuses from military to civilian producers. Even when divergent specifications are not a problem, generic military-sponsored technology with commercial applications may simply be too slow to diffuse to civilian producers—particularly in instances when alternative civilian applications of the same underlying technology have already appeared on the market.

As the previous chapters have demonstrated, this latter scenario is increasingly likely to be the case. Even more disturbing for those concerned with American security are the military uses to which such foreign-born commercial technology can increasingly be put. Many of America's most vaunted weapons—from the AMRAAM missile to the M-1 tank—literally could not be built without commercially developed Japanese machine tools. Without commercially derived Japanese components for their radars, America's F-16 fighter pilots could never find their targets. Regardless of its impact on the nation's commercial competitiveness, the military's traditional approach to developing advanced technology may already be obsolete for its own purposes.

Yesterday's Spin-Offs: Both Sides of the Story

Case studies of previous Pentagon attempts to involve commercial firms in military technology development are a useful way to gauge the potential of a spin-on strategy against the spin-off strategy that has characterized U.S. technology policy since the end of World War II. Looking back over the years, it is evident that some of these attempts have been successful, and some have not. In the 1950s and early 1960s, the competitive position of civilian industry was often strengthened by military-oriented technology projects. In other instances, the commercial impact of military projects was decidedly negative. Since the late 1970s the impact of military projects on commercial competitiveness has been at best negligible. By examining cases from both periods, we can identify the circumstances that have produced each outcome.

Case Study 1: Solid-State Transistors and Integrated Circuits

The development of solid-state microelectronics—transistors and integrated circuits—is cited often as a case of positive commercial spin-off from the military sector, and with good reason. It is important to note, however, that even at its most successful, spin-off still represented a second-best solution to the problem of promoting the commercial technology base. As early as 1958, the military's emphasis on military-specific devices conflicted with the commercial interests of the Bell System, causing Bell executives to worry that military design specifications might undermine the production efficiency of AT&T's manufacturing arm, Western Electric.

Nevertheless, military procurement provided a crucial launch market for the untried semiconductor technology, fostering a market environment that encouraged entrepreneurial risk taking. Concerned that the military might classify its technological breakthroughs, Bell Telephone Laboratories rushed to make its semiconductor innovations public and its patents marketable. The Pentagon also required a domestic second source for its semiconductor purchases, a requirement that further lowered market-entry barriers and accelerated the diffusion of technological advances. Thus, the early, military-structured market environment clearly promoted the development of a strong, independent semiconductor industry in the United States. The differences between this case and the case of numerical controls (which follows) illustrate the key role played by the technical requirements of the highest-volume user—then and now—in determining the prospects for successful spin-offs.

Nineteen hundred forty-eight had witnessed a fortuitous match of military needs and commercial objectives. Driven by the rigorous performance requirements of its military mission, the Army Signal Corps had established a goal of miniaturizing electronic communications gear. At the same time, scientists at Bell Labs were searching for an effective solid-state amplifying device to replace mechanical relays in telephone exchanges. Both objectives were initially satisfied by the invention of the point-contact transistor, announced by Bell Labs in 1948.

The Army Signal Corps's effort to miniaturize military electronics had begun in the late 1930s and culminated in the first "walkie-talkie." Although it represented a major improvement in battlefield communications, the six-pound, football-sized, two-vacuum tube transmitter-receiver with separate telephone handpiece was too bulky for many military operations. Miniaturization thus received special emphasis in the Signal Corps's long-range R&D plans after World War II. Working with the electronics industry, the Signal Corps soon developed an automatic soldering system (Auto-Sembly) to facilitate the mass production of miniaturized components.[2]

One week before the first public demonstration of the transistor, Bell Labs held a special briefing for the military services. Bell scientists had long been aware of the military's interest in their work. They waited as long as possible to disclose their results for fear that disclosure of the transistor to the military prior to a public announcement would lead to severe restrictions on its commercial use or outright classification in the name of national security.[3]

As expected, researchers from the Signal Corps Labs were immediately enthusiastic about the new device, sensing its amenability to the Auto-Sembly technique. Within months, the Corps had set up a small manufacturing facility to produce test devices. By June 1949, the military had convinced Bell to sign a contract for the study of potential applications. This contract eventually resulted in the first published research on the applications of transistors to digital computers.

Although this first military contract supported general application and circuit studies, Bell's second military contract specified that materials, services, and facilities be devoted to studies of military interest. Internally, Bell began to coordinate its transistor development with military requirements. The military market was important because the Bell System was having trouble introducing transistors into the telephone system. The phone system could only introduce transistors as older vacuum tube equipment was retired; this process was to take more than a decade. Also limiting nonmilitary markets at first was the high cost of solid-state devices. The first transistorized hearing aid sold for $229.50. Raytheon's 1955 transistor radio was considered a luxury item and retailed for $80.[4] Unlike radio listeners and hearing aid users, the military could subsidize the technology's development costs in order to increase the scale of production.

Military money underwrote the construction of a huge Western Electric transistor plant in Pennsylvania; Raytheon, RCA, and GE also benefited from military support. In return for guaranteed government purchases (at premium prices) of a part of their output, the companies agreed to build production capacity ten to twelve times greater than that needed to supply the government's existing demands. The government's request apparently related to the military's constant interest in "surge capability," that is, the capacity to ramp up expanded production rapidly in case of a wartime emergency.[5] In peacetime, however, the resulting excess production capacity created a further incentive for the industry to develop new commercial markets.[6]

In addition to subsidizing plant construction and enlarging production capacity, military contracts facilitated the dissemination of information about the underlying technology to potential users outside of the military services. As required by its second military transistor contract, Bell Labs held a symposium on transistor applications in September 1951. The lectures and demonstrations were attended by over 300 representatives of academia, the military services, and the electronics industry. Each participant received a 792-page volume of the symposium proceedings; in addition, the military services distributed 5,500 copies at government expense. A second conference (in April 1952), funded by Western Electric but held at the urging of the military services, resulted in two fat volumes detailing the scientific fundamentals of transistor technology. Thus, at a time when Western Electric, AT&T's manufacturing arm, was only just beginning transistor production for trial use in AT&T's civilian long-distance telephone system, Bell Labs was already disseminating the transistor technology that "enabled all [Western Electric] licensees to get into the military contracting business quickly and soundly."[7]

Military support for transistor research at Bell rose from a small level in 1950 to 50 percent between 1953 and 1955. By the mid-1950s, Bell Labs scientists had made two major technological advances. Advances in solid-state theory and metallurgical techniques made possible the creation of "junction" transistors, which were mechanically less fragile than point-contact transistors. In addition, the invention of the diffusion technique for manufacturing transistors made possible the mass production of devices that could amplify high frequencies.

As basic transistor development proceeded, however, the applications desired by military users began to diverge more and more from the types of applications that Bell Labs scientists envisioned for use in the public telephone network. Aside from bulkiness, additional problems with military communications equipment had been revealed during the war. Walkie-talkies had failed when confronted with extremes of temperature—freezing in the Arctic, moisture and corrosion in the jungle. They had also failed when subjected to other rigors of battlefield use, including shock, vibration, and sudden changes in temperature. These problems led the military services to prefer devices constructed from silicon instead of germanium. The ambient temperature in jet aircraft and guided missiles, for instance, often exceeded the 75 C maximum operating temperature of germanium transistors; silicon devices would continue to operate in environments as hot as 150 C. The expansion of the Navy's nuclear fleet and the Air Force's plan to develop a nuclear-powered plane also increased the importance of silicon's inherent resistance to radiation.

The U.S. military's interest in silicon transistors set the American industry on a development trajectory decidedly different from the germanium-based development trajectories that were being explored in Europe and Japan in the 1950s. That interest also influenced the structure of the emerging U.S. semiconductor industry: It enabled the primary manufacturer of silicon transistors, Texas Instruments, to carve out an early niche in the emerging semiconductor market.[8]

In addition to promoting silicon over germanium, military requirements fostered attempts to exploit certain physical properties of semiconductors that were not likely to find wide commercial application. As important as miniaturization, heat resistance, and resistance to radiation all were for military users, for example, what the Signal Corps most wanted was a transistor capable of amplifying very high frequency (VHF) signals for computers and communications equipment. This led the military services to prefer diffused or "intrinsic barrier" devices to the point-contact and junction devices most commonly used in Bell System applications. (Bell researchers had invented diffused devices in 1954 in a project supported by a Signal Corps contract.) Intrinsic barrier devices were difficult and expensive to manufacture, but they were used in several military applications. Bell needed only small numbers of diffused germanium devices, but missiles required high-switching speeds and so the military demanded large numbers of diffused transistors, mostly silicon.

Flush with new R&D funds due to the massive rearmament effort that followed the Korean war, the Signal Corps actively pushed the electronics industry to provide transistors in the form that it wanted. In 1956 the Corps placed $15 million worth of development contracts for work on diffused devices, the largest amount of R&D funding ever awarded until that time. The stated purpose of these awards was to "make available to military users new devices capable of operating in the very high frequency (VHF) range which was of particular interest to the Signal Corps communications program.[9]

Although the awards provided the military services with the devices they wanted, it became increasingly clear that the trajectory of commercial development was headed in a different direction. By the end of 1958, Western Electric had manufactured 171,000 diffused transistors. All of them were destined for military applications and none for internal consumption by the Bell System. Internal memoranda circulating around Bell Labs at the time suggested that big military projects were taking too many Bell engineers away from nonmilitary development work. What is more, Bell telephone network applications were beginning to outsell military applications—two large telephone projects accounted for over 1 million transistor sales in 1956, when total military sales were expected to be only 175,000.[10]

Still, the military services continued to affect the specific forms that the developing technology would take. A rapid proliferation of transistor types appeared in the 1950s as different firms investigated various potential applications (and different development trajectories). Because it was too hard to integrate several transistor types into a single system, the military pushed for standardization. The Signal Corps had already sponsored a conference in mid-1953 aimed at standardizing the operating characteristics of transistors. (By contrast, the British semiconductor industry remained without national standards until the 1960s.)[11] In 1957, the head of the Bell Labs Nike missile development project came up with a plan to standardize and expedite the development of transistors needed specifically for military systems. The so-called preferred devices—both germanium transistors for high-frequency needs and silicon to meet high temperature requirements—were all diffused.[12]

Military spending on microelectronics research continued to expand into the 1960s, but it was military procurement that did the most to shape both semiconductor technology and the entrepreneurial dynamic of the American semiconductor industry. During the 1950s the military services had supported a number of research projects aimed at the development of new microelectronic components for the next generation of military weaponry (radar, fire control systems, missile guidance systems). All of these projects, including Tinkertoy and Micromodule, were dominated by established electronics firms, such as General Electric, Hughes, and RCA; none of them were successful.[13] The first integrated circuit was eventually demonstrated in 1958 by Texas Instruments (TI), which had developed its device without direct research and development support from the military. Nevertheless, Jack Kilby's own account of his invention confirms that Texas Instruments had solely military applications in mind when research into the new devices began.[14]

Once the integrated circuit had become a reality, the armed forces spared no effort to support its further development. Between 1959 and 1962, TI, Westinghouse, and Motorola received $9 million worth of military contracts for further work on integrated circuits. TI alone received a $1.15 million, two-and-a-half-year development contract in mid-1959, followed by an additional $2.1 million contract, awarded at the end of 1960, to come up with special manufacturing equipment and production techniques that would allow the new devices to be mass produced.[15] Despite the military's early enthusiasm, however, commercial producers remained wary. As late as 1961, many scientists in industry and academia still voiced doubts that integrated circuits would work in actual electronic equipment and systems.[16] Having decided in 1958 to employ integrated circuits in the guidance system of the Minuteman missile, the Air Force was concerned to alleviate these doubts once and for all. Under Air Force sponsorship, a small digital computer was introduced into the TI production program. Two demonstration computers were built—one was made with discrete semiconductors and required 9,000 individual components; the other performed identically, but contained only 587 integrated circuits.

By providing an initial market at premium prices for major advances, military purchasers accelerated their introduction into use.[17] As production for the military proceeded, producers accumulated experience. Experience soon translated into lower unit costs. Within a few years, the price of a typical device was low enough to spawn civilian applications, first in industry, then in consumer products. Nevertheless, the most significant consequence of early military interest in integrated circuits occurred on the supply side. In the era of the vacuum tube, the manufacture of electronic components was dominated by large, multi-divisional producers of electronic systems; as semiconductor technology gained ground, this continued to be the case in Western Europe and Japan. In the United States, by contrast, military procurement decisions effectively created an independent segment of "merchant" semiconductor suppliers—that is, companies organized to manufacture semiconductors primarily for sale on the open market, instead of primarily for internal use.[18]

As early as 1959, new merchant producers accounted for 69 percent of all semiconductor sales to the military services (a figure that translated to 63 percent of total semiconductor sales). Texas Instruments and Fairchild, the two new firms that had pioneered the development of integrated circuits (ICs), became major military subcontractors. TI was charged with providing integrated circuits for the Minuteman II missile guidance system, and Fairchild got the NASA contract to provide an IC-based guidance computer for the Apollo spacecraft. The rapidly increasing utilization of integrated circuits by both NASA and the U.S. Department of Defense (DOD) enabled other emerging suppliers, such as Motorola and Signetics, to intensify their focus on IC production.

Besides its contribution to the creation of a flexible and highly independent production network for semiconductor components, military policy also encouraged the widespread dissemination of basic technological information among competing companies. The DOD typically obtained a comprehensive license of free use for any patentable products or processes that had been developed by private contractors with military funds (and for military use). This meant that firms that had developed commercially relevant innovations in the context of a military project had to share information about the innovation with any firm or individual who subsequently worked on a government contract or a private project supported by government funds.[19] Moreover, the DOD typically followed a strategy of "second-sourcing," that is, requiring at least two independent sources for a component before it could be included on an approved list for use in military equipment. Second-sourcing also encouraged rapid technological diffusion; co-contractors often shared patent rights, drawings, photomasks, and manufacturing know-how.[20]

In sum, instead of yoking suppliers to military users and thus privileging the development of military-specific forms of IC technology, military policy contributed to the development of an industrial structure that would soon prove highly beneficial to the rapid commercialization of IC technology. The very independence of the semiconductor merchants encouraged them, indeed required them, to pursue all possible alternative applications of the underlying technology. As new companies entered the industry (Signetics, Silconix, General Microelectronics, Molectro), total IC production mushroomed, growing from about $4 million in 1963 to roughly $80 million in 1965. IC prices fell (from $31 in 1963 to below $9 by 1965), emboldening older producers of electronic systems (RCA, Sylvania, Motorola, Raytheon, Westinghouse) to move into volume production.[21] As civilian computer and industrial applications increased in the mid-1960s, the importance and influence of military procurement declined rapidly. Government (mostly military) users accounted for 100 percent of the U.S. market for ICs in 1962, 55 percent in 1965, and 36 percent in 1969. By 1978, government's share of the American IC market hovered around 10 percent.[22]

Case Study 2: Numerical Controls

Although some tension had developed between military-specific performance requirements and the commercial interests of fledgling semiconductor producers (particularly Bell Labs), military procurement policy—price subsidies, second-sourcing—had simultaneously fostered the development of an entrepreneurial industry and a commercial market that soon outpaced the military market in both size and technological sophistication. Spin-off was successful because Pentagon policy ultimately allowed the technical needs of high-volume commercial users to dominate the market and drive the technology.

Military intervention did not always have such salutary commercial effects, however, even in the heyday of successful spin-offs. Military performance requirements and procurement practices actually impeded the civilian diffusion of innovative computer control technology for machine tools. The commercial industry's overreliance on Pentagon-sponsored technology and funding—an overreliance on military spin-offs—ultimately created a domestic military-industrial enclave. The enclave came to be inhabited by firms whose market antennae and technological requirements were fine-tuned to the specialized needs of military users. This rendered them increasingly out of tune with the needs of a (potentially) much larger civilian market.

Between 1949 and 1959, when the Air Force discontinued its formal support of software development, the military spent at least $62 million to research, develop, and diffuse numerical control technology, most of it originating at MIT's Servomechanism Laboratory.[23] In this initial phase of the technology's evolution, Air Force performance specifications defined a unique development trajectory for Servo Lab engineers, a trajectory linked to end-use requirements that were well beyond the standard needs of most potential commercial users. Specific performance requirements for four- and five-axis machining stemmed from the Air Force's need to fashion large, structurally complex metal parts (integrally stiffened wing sections, variable-thickness skins, etc.) out of tricky materials, as components for high-speed aircraft and missiles.[24] The Air Force also wanted a system that could rapidly be reprogrammed in an emergency over commercial communication channels. Together these requirements led Servo Lab engineers to create a software system known as APT (Automatically Programmed Tools) that was at once universally applicable and infinitely adaptable, with the additional ability to control the motions of a cutting tool along five axes in unbounded space.[25]

The same technical characteristics that made this trajectory so appealing to military users rendered the system overly expensive and sophisticated for most commercial applications. The dominant programming approach followed by early industrial users of numerical control involved the development of a two-dimensional executive routine, or "pilot program." The pilot program could be programmed to select from a library of subroutines, coordinating them for more complex three-dimensional work. APT transcended the coordination problem by creating a more fundamental system that recognized general categories of cutting problems which could then be particularized for individual surfaces and dimensions. The fact that APT was more fundamental, more flexible and adaptable, and more capable of growth than the subroutine approach also meant that it was more cumbersome and more prone to error; APT required the largest available computers and the most highly skilled mathematicians to program them.[26] In brief, the system possessed all of the features required to meet the specialized demands of the Air Force and "none of the generality and simplicity required to express economically and in an easily learned way the huge range of everyday machine operations."[27]

A military-specific development trajectory for numerical control (NC) technology pervaded the American machine tool industry on the demand side because large-scale military development efforts occurred during the period of the technology's initial creation, when interest among commercial firms was distinctly cool. The entrenchment of that trajectory, or the creation of a military-dominated trajectory among the industry's core firms, occurred on the supply side, in the subsequent organization of the supplier network and its characteristic relationships with large machine tool users. The argument, in brief, is that military involvement promoted the creation of a specialized production infrastructure for NC tools. A combination of industry structure and military policy turned the attention of NC tool suppliers away from the work of developing potential applications for high-volume producers of consumer durables.

Between 1949 and 1953, MIT and the Air Force both mounted massive campaigns to interest commercial machine tool builders in NC; in all that time, however, only one private company, Giddings and Lewis, Inc., was interested enough in the new technology to invest even a portion of its own funds. Partly in response, the Air Force set about creating both a market and a set of preferred suppliers for NCs. The Air Force paid for the purchase, installation, and maintenance of over 100 NC machines in prime contractors' (mostly aerospace) factories and funded training programs to teach the contractors how to use the new technology. In 1955, promoters of the technology successfully changed the specifications for stockpiling machine tools in the Air Material Command budget allocation from tracer-controlled to numerically controlled machines.[28] The results were impressive: Between 1951 and 1957, private research and development expenditures in the U.S. machine tool industry multiplied eightfold, most of it underwritten by the aerospace industry.[29]

The Air Force practice of "seeding"—that is, placing NC tools with selected users—provided these large-scale, technologically advanced firms with a dominant hold on the NC segment of the machine tool market.[30] Combined with the high price and technical complexity of the tools, Air Force contracts worked to restrict the market to the aerospace industry and similar specialized uses. In the absence of an industrial policy that might have identified and perhaps even subsidized commercial alternatives, the industry's emphasis on meeting military-related needs made sense from a business perspective.[31] High-level aircraft industry executives desired cost-plus contracts with the Air Force and overcame initial resistance within their companies to the exclusive use of APT.[32]

Beyond making APT the industry standard, military policy actively narrowed the diffusion of the latest APT developments to its own clients, after which the developments became in effect proprietary information that could be used to commercial advantage. Ongoing APT research was shifted to the Research Institute of the Illinois Institute of Technology (IITRI) in 1961, where it has been guided by a consortium composed of the Air Force, the Aircraft Industries Association (AIA), and major suppliers of machine tools and electronic controls. Membership in the consortium is expensive, beyond the financial means of most companies in the metalworking industry. Access to APT systems has thus been effectively restricted to AIA members (including Boeing, Lockheed, Convair, Chance Vought, Bell, Martin, McDonnell Douglas, North American Aviation, Northrop, Republic, and United Aircraft) and affluent nonmembers such as General Motors, Goodyear, IBM, and Union Carbide. As late as 1983, APT information was treated as proprietary within user plants; programmers were required to sign out manuals for use at the plant (they could not take them home), and they were forbidden from discussing the technology with people outside the company.[33]

Over time, the military-dependent environment fostered close collaborative ties between large users and the largest machine tool suppliers, further promoting specialized end-use developments and the establishment of advanced manufacturing systems geared toward the particular needs of large aerospace firms. The standardization around APT, promoted by the Air Force, inhibited for more than a decade the development of simpler programming languages that might have made contour programming more accessible to smaller machine shops. When breakthroughs in programming methods (for example, the creation of Compact II by Manufacturing Data Systems, Inc. (MDSI)) and computer design (for example, the invention of microprocessors) made it possible to develop low-cost, mass-produced NC tools in the 1970s, most large American tool makers simply had too little experience with consumer durables producers to detect and exploit the unmet commercial demand for such equipment. Smaller tool suppliers that did attempt to adopt numerical controls for commercial applications were forced into a position of technical dependence on the large firms that had controlled the development of APT.

By contrast, Japanese machine tool makers benefited from their close ties to major *industrial* users and were able to shift more rapidly to the mass production of smaller NC equipment in the early 1970s. Many Japanese machine tool firms are owned wholly or partly by their major industrial customers; Toyota Koki is partly owned, for instance, by the large automobile manufacturer Toyota. According to Friedman (1988), many small and medium-sized tool producers are also linked to large industrial users through geographic concentration; within these clusters, groups of tool firms specialize in tools needed for specific industries and benefit from long-term production and subcontracting arrangements. These ties have presented Japanese tool makers with an opportunity to develop NC technology with an array of commercial end uses in mind. Aided with finances and technical expertise from other divisions of their corporate groups, Japanese tool producers were quicker than their American and European counterparts to introduce microprocessors into their

NC systems in the early 1970s.[34] In all, Japanese production of low-cost numerically controlled lathes and machining centers increased tenfold between 1970 and 1979.[35]

Although Japanese successes in this sector were due in part to government policy, for the most part, it was the Japanese tool makers' close ties to large industrial users that provided them with an early and consistent source of alternative development trajectories for NC technology.[36] MITI sponsored the rapid diffusion of NC technology throughout the Japanese economy, operating through a set of regional technical assistance centers—financed from bets collected on company-sponsored bicycle races—to teach small and medium-sized metalworking shops how to use NC tools.[37] But private Japanese firms began as early as 1955 to apply work done by MIT's Servo Lab engineers, and the first computer-controlled NC tools developed in Japan were shown at an international exposition in 1958.[38]

By the late 1970s, market share figures were revealing the damaging consequences for U.S. competitiveness of a military-dominated development trajectory for the technology of numerical control. By 1979, more than two decades after the technology became commercially available, only 2 percent of all machine tools used in the United States were numerically controlled; in 1978, only 3.7 percent of the metalworking equipment used by the U.S. machine tool industry itself was numerically controlled.[39] Although the total number of NC machine tools almost doubled between 1978 and 1982, imports as a share of the value of U.S. consumption rose from a little over 23 percent in 1980 to more than 35 percent by 1983, almost 90 percent of them from Japan.[40] In 1984, two-thirds of the numerically controlled turning machines and three-quarters of the NC machining centers installed in U.S. firms were bought from foreign firms.[41] During the first seven months of 1985, more than 50 percent of all NC tools used in the United States came from overseas.[42] The fall of the dollar after early 1985 slowed the competitive decline of U.S. producers somewhat, but the industry's long-term prospects are still considered precarious.

Spin-Off Today: The Dilemmas of "Dual Use"

In the late 1970s, the spread of America's competitive troubles to such high-tech sectors as semiconductors and computer-controlled machine tools was still typically linked to Japanese protectionism rather than to a decline in native technological prowess. Still, many analysts found the trend unsettling. Toward the end of the Carter Administration, a series of apprehensive research reports appeared from various outposts of the military establishment.[43] The reports decried the deteriorating state of the nation's "defense industrial base" (DIB), the industrial infrastructure that formed, according to proponents, the civilian backbone of the nation's military posture.[44] In response, military planners began to work with their consultants in private industry to devise technology development projects that would involve commercial firms in planning, promote the use of commercially available components (whenever possible), and subsidize the commercial development of basic "dual-use" technologies that were expected to find wide application in both the military and commercial sectors. Two Pentagon projects, the Very High Speed Integrated Circuits (VHSIC) program and the Strategic Computing Program (SCP), are illustrative of this approach. Both ran throughout much of the 1980s and have had a negligible impact on the commercialization of their constituent technologies.

Unlike the Pentagon's earlier semiconductor and numerical control efforts, VHSIC and SCP occurred when a set of commercial development trajectories for semiconductor and computing technologies were already well established, not only in the United States but also in Western Europe and Japan. In an environment where multiple-development trajectories already existed, military-specific end uses rapidly became the *raison d'etre* for the military projects. Despite their best efforts to promote "dual use," Pentagon planners were ultimately (and perhaps appropriately) most interested in inducing leading-edge commercial producers to supply military applications that none of their commercial customers wanted.

Case Study 3: VHSIC

In early 1977, President Jimmy Carter, Secretary of Defense Harold Brown, and Brown's director of research, William Perry, agreed to push for the rapid development and deployment of the cruise missile. Characterized by its reliance on state-of-the-art electronics to steer close to the ground—thereby avoiding enemy radar—the cruise missile (like other precision-guided munitions) epitomized the strategic use of U.S. technology to offset Soviet numerical superiority in conventional long-range weaponry. Nevertheless, the cruise missile's apparent dependence on research breakthroughs in microelectronics and materials science soon underscored a concern that had already been growing within the Pentagon about the so-called insertion lag. The term refers to the amount of time that elapses between the commercial availability of an integrated circuit and its utilization in a weapons system. By the late 1970s, the insertion lag reportedly had reached a span of twelve years. An intelligence report in the fall of 1977 suggested, further, that the U.S. military's lead over the Soviets in microelectronics had dwindled to substantially less time than that.[45]

Through the rest of 1977 and 1978, Perry and a handful of DOD officials set about establishing a major university-based research program in very high speed integrated circuits (VHSIC), solid-state components that are essential to the operation of cruise missiles and other precision-guided munitions. VHSIC was to emphasize advanced computer and data-processing architecture, new approaches to computer-aided design (CAD) of complex circuits, and research into the materials and physical processes needed to achieve submicron geometries—all areas at the cutting edge of technological development. The program was in many ways a response to the demands of military users who had become increasingly frustrated over their inability to convince commercial semiconductor producers to develop custom chips for military applications.[46] With less than 10 percent of the semiconductor market (compared to more than 50 percent through the mid-1960s), producers of military systems had little financial leverage left with which to entice innovative semiconductor firms into doing military contract work.[47] "We were forced to use decade-old microelectronic technology," complained one Pentagon official, "while Atari games were using the latest."[48]

Commercial semiconductor producers understandably preferred to focus their financial and design resources on commodity chips that were likely to command lucrative large markets, such as those used in personal computers and video games.[49] By the mid-1970s, commercial semiconductor technology had advanced several years ahead of concurrent developments in the military sector, and cutting-edge developments were clearly being driven by the needs of large commercial users.

Appearing before the Senate Armed Services Committee in 1979, Perry made no secret of the Pentagon's desire to regain control of the technological agenda in microelectronics by promoting the concept of dual use. VHSIC's planners intended to "direct the next generation of large-scale integrated circuits to those characteristics most significant to defense applications," he testified. VHSIC would, furthermore, "insure that the U.S. maintains a commanding lead in semiconductor technology and that this technology will achieve its full potential in our next generation of weapons systems."[50]

With the merchant semiconductor producers' research and production efforts geared overwhelmingly toward industrial and consumer markets, Pentagon planners recognized that merchant decisions to participate in VHSIC would be treated as a matter of corporate strategy, not national security. Each firm's decision to participate would be based primarily on a belief that VHSIC objectives matched the objectives of in-house technology development, and a conviction that participation in the program would accelerate in-house progress toward meeting common developmental goals.[51] An important, though secondary, spur to participation was the notion that VHSIC participants would be the first to profit from any commercial spillovers that might arise from VHSIC technology.[52]

Spillovers and military-civilian complementarity were the core objectives of a dual-use strategy. Indeed, the identification of potential spillovers and complementarities was a major goal of VHSIC planners, who recognized early on that "VHSIC could not be sold to industry on national interest alone or even in large part."[53] An early planning document stated that VHSIC technologies "must be consistent with mainstream industry efforts" and that "program goals must be consistent with the industry learning process."[54] Following the first rule of a dual-use strategy, VHSIC planners sought and incorporated substantial advice from private industry about appropriate technical goals and organizational features for the program.

In particular, the Pentagon's Advisory Group on Electron Devices (AGED) provided a critical institutional link between government and private industry during VHSIC's planning.[55] Composed of industrial and academic specialists, AGED sponsored two Special Technology Area Reviews in September and November 1978. Thirteen companies were invited to make formal presentations on prospective VHSIC technology: Fairchild, GE, Hughes, IBM, Motorola, National Semiconductor, Raytheon, RCA, Rockwell, Sandia, Texas Instruments, TRW, and Westinghouse.[56]

Concerned to rein in military-oriented technical goals that far outstripped the near-term needs of commercial users, industry representatives convinced VHSIC planners to focus solely on silicon devices, rather than on both silicon and gallium arsenide. They also convinced them to scale back the program's interim goal for feature size from submicron line widths to a more reasonable width of 1.25 microns after the program's first three years. Industry input was also instrumental in increasing the program's development support for CAD tools.

In addition to affecting the program's technical end-use requirements, AGED reviewers addressed the area of supply-side organization. The reviewers evolved the notion of multifirm "teaming." The idea of teaming VHSIC users and suppliers was a departure from the Pentagon's usual practice of making awards to numerous individual contractors. Though substantial consultation had occurred in the past between the military equipment and military components divisions of companies working on procurement contracts, collaboration between the R&D divisions of different firms was not a widespread practice. AGED reviewers thought this should change. In order to improve communication between systems experts in user firms and components experts in supplier firms, companies were asked to apply for VHSIC contracts in teams. Each team would include the manufacturer of a military system and a merchant semiconductor firm, plus companies specializing in other technically relevant areas such as design, processing, packaging, and testing.

Participating semiconductor firms expected to benefit commercially from anticipated advances in the areas of CAD and lithography. CAD techniques seemed especially suited to the strategic needs of commercial producers, since their wider introduction would have the effect of "shifting some of the design burden from the device manufacturer to the user, [thus allowing] semiconductor companies to reduce the amount of engineering time they must commit to new product development."[57] Commercial spillovers were anticipated primarily in the area of process technology. The development of a CMOS (complementary metal oxide semiconductor) production line for VHSIC circuits was expected, for example, to advance the process technology for other CMOS applications in the fields of microprocessors and telecommunications.[58]

In practice, the notion of complementarity between civilian and military goals was manifested in the way most participating firms organized their VHSIC work. In most cases, VHSIC activities were fused with each firm's mainstream development and production efforts, rather than being isolated in separate military divisions. For example, VHSIC work at Motorola, National Semiconductor, and Texas Instruments proceeded under the direction of the same corporate vice presidents who were responsible for overall semiconductor R&D. Texas Instruments did not separate VHSIC and commercial Very Large Scale Integration (VLSI) work at all, and VHSIC engineers at Motorola and National were drawn from, or worked closely with, engineering personnel who had been involved in each company's ongoing VLSI efforts.[59]

Yet the project's efforts to promote the idea of dual use seemed to bear little fruit; only one VHSIC chip had actually been built into a military system by the end of 1987.[60] This means that, at best, the "insertion lag" had been reduced from twelve to seven years. Honeywell, a major VHSIC contractor, moved its chips from production on six-inch diameter silicon wafers to a four-inch diameter line, citing lagging demand for Phase I chips. What is more, many *commercially* developed chips from the VLSI generation could meet all of the technical parameters that were said to define military-sponsored VHSIC chips (a geometry of 1.25 microns or less, ability to run at 25 megahertz or more, with a total density of 500 billion gate-hertz per square centimeter).[61]

Despite the attention of program planners to the common needs of military and commercial users, three military objectives actually came to dominate the technological agenda of VHSIC participants. First, VHSIC was designed to support the development of advanced integrated circuits for incorporation into specific military systems. Second, the program was to promote the introduction of such circuits into military systems in a "timely and affordable" manner. Third, the program was supposed to ensure that American technology would surpass any technologies potentially available to the Soviets for use in *their* advanced weapons systems. Indeed, technology managers in at least one electronics firm used the firm's participation in VHSIC internally to justify continued high spending on commercial R&D; because of the technological complementarities involved, it was argued that one would be substantially wasted without the other.[62]

These objectives influenced both the direction of technological development and the organization of the production infrastructure that ultimately carried it out. The program's technical specifications created a military-oriented technology trajectory that diverged from the needs of most commercial users. This military orientation was then entrenched by the creation of a dedicated production network and by military policies that inhibited the commercial diffusion of VHSIC technology. Nevertheless, in an environment where multiple commercial development trajectories already existed for the underlying technology, this supply-side entrenchment did no particular damage to the technology's prospects for commercialization. Instead, VHSIC's demand-driven military trajectory rendered the program virtually irrelevant to the commercial sector.[63]

Although it may seem ironic from the perspective of those who favor a dual-use development strategy, the inclusion of private industry representatives in the planning stages of VHSIC actually led program managers to emphasize end-use specifications that diverged from the needs of most commercial users. Drawing on a wealth of technical experience, AGED reviewers understood from the beginning that system considerations would have to guide the design of VHSIC components. VHSIC planners had expected the program to focus on the development of specific devices, emphasizing smaller features and faster speeds. AGED reviewers advised them, however, that it would be highly inefficient to develop high-speed circuits in isolation from the electronic systems that would later have to incorporate them. Consequently, VHSIC planners stipulated that circuit designs would be system-driven, that is, designed to meet the specific needs and requirements of military hardware.

Military performance specifications thus defined a unique development trajectory for all program participants. The systems orientation soon encompassed all VHSIC-sponsored work, causing engineers to make technical choices that diverged from the priorities of commercial users. For example, in order to link chip designs to the requirements of particular weapons systems, five of the six Phase I teams developed custom or application-specific chips. Only Texas Instruments, which combined its own systems and semiconductor divisions to form a single VHSIC team, created a standard chip that could be programmed to adapt to various military and civilian applications.[64] As for those teams that took the custom design route, military requirements again skewed technological choices away from paths more relevant to commercial needs. For example, VHSIC contractors met the Pentagon's second goal of minimizing turnaround time between technology development and insertion into final systems by choosing design tools that did not maximize utilization of the chip's surface. For profit-seeking commercial firms, however, cost is directly related to circuit density per chip; the VHSIC program deflected these firms from seeking high-density designs that might have paid off over long production runs.[65]

This military-oriented development trajectory was soon entrenched among VHSIC participants by the creation of a VHSIC-specific network of users and producers. Competition was encouraged by forcing VHSIC teams to bid on each of three contract phases; the competition narrowed with each phase. In each case, however, the range of possible bids was already constrained to meet the specific requirements of military systems. Six VHSIC Phase I contractors set up special production lines; of the six, only one, Texas Instruments, primarily served the civilian semiconductor market.[66]

Rather than create a dedicated network of VHSIC contractors, the Pentagon might have chosen to reform its procurement practices. This would have made it easier for private firms to develop VHSIC components that were commercially viable and simultaneously attractive to the DOD (that is, military spin-on instead of commercial spin-off). DOD officials reasoned, however, that a dedicated network would be more responsive to military needs.[67] Moreover, only by creating a dedicated production network under Pentagon control could VHSIC officials be certain that they were getting components that were better than—or at least, different from—components that the Soviets might potentially obtain in international markets.[68]

From the very beginning, similar concerns over national security had precipitated a serious conflict between VHSIC planners and other groups in and around the Pentagon. Worried that VHSIC technology might find its way too easily into the hands of commercial (and, ultimately, military) competitors, the House Armed Services Committee decided (with the support of various Pentagon officials) to place strict export controls on any "technical data" developed in the course of VHSIC research. Under the provisions of the International Traffic in Arms Regulations (ITAR) passed by Congress as part of the Arms Export Control Act of 1976, government permission would be required for the export of any VHSIC technology or technical data subject to potential military application.

In line with the congressional restrictions, VHSIC's director issued a memorandum to each of the armed forces in December 1980 stating that VHSIC research would be subject to export controls henceforth under both ITAR (which is administered by the State Department) and the Export Administration Regulations (EAR) which are administered by the U.S. Department of Commerce. Although these regulations were meant to apply primarily to VHSIC prime contractors and their semiconductor suppliers, university research was also included, given the practical impossibility of distinguishing basic research from process technology with potential military applications. According to a report jointly commissioned by Perry's office, DARPA, and the Office of International Security Affairs at the U.S. Department of Energy, such controls might apply even to the mere communication of basic research results to foreign scientists, even within the United States.[69]

Indeed, restrictions were placed on discussion in open (nonclassified) technical symposia of either the architectures or performance characteristics of VHSIC contractors' chips; manufacturers were forbidden from discussing details of the software used in either their CAD or their fabrication processes. Pentagon restrictions became so stringent, in fact, that VHSIC contractors were forbidden even to publish close-up, front-forward photographs of their products. One story had it that, when the U.S. General Accounting Office insisted on such photos for an unclassified report on the program, VHSIC officials were directed to send, instead, an aerial photograph of a parking lot, reduced in size until the cars resembled a cluster of microcircuitry.[70]

Over time, the fear that export and publication restrictions might spill over onto their commercial operations led some VHSIC contractors to isolate their military work from internally funded commercial R&D.[71] Brueckner and Borrus (1984) interviewed the manager of the commercial Large Scale Integration (LSI) division of a large VHSIC prime contractor who "indicated that he keeps "copious files" detailing the complete abyss between VHSIC and VLSI research" even though the firm's commercial signal-processing components were "extremely similar" to VHSIC circuits. The manager indicated that the company was pursuing parallel research efforts in order not to subject *commercial* research and products to DOD publication and export controls.[72]

Many VHSIC participants continued to argue, nevertheless, that relevant VHSIC technology inevitably would find its way into commercial semiconductor products: "If we are using improved fabrication techniques in one part of our manufacturing facilities to produce VHSIC devices, and these techniques are sorely needed in our commercial device facility to meet Japanese competition, then the VHSIC technology will `diffuse' into our commercial operations in spite of Pentagon-imposed barriers."[73] Worries over restrictions on commercial diffusion may have been misplaced, however, since commercial processing technology has kept pace—indeed, has often outpaced—technology developed during the course of the VHSIC project. By 1988, VHSIC had met its Phase I goals, establishing facilities that could build chips with geometries 1.25 microns apart. By then, however, commercial chips were already being built with geometries as fine as 0.7 and 0.8 microns. Moreover, instead of working to refine existing optical lithography techniques to achieve the smaller device geometries, VHSIC contractors "brute-forced" it with expensive E-beam and X-ray lithography. The task of pushing less expensive optical lithography into the submicron range was left to commercial chip producers, which would have followed the same course had VHSIC never existed.[74] Asked late in 1986 whether commercial applications would grow out of VHSIC technology, the program's director was unusually candid:

> That may happen, but my answer to that question is that if we did our job very, very well, there shouldn't be any commercial applications for these things. That is what we are supposed to be working on—the part that the commercial market wouldn't serve and the military needs. Nobody will build these kinds of chips because there is no way to recover their money in the commercial market, so we have to pay to get them designed. But engineers are bright people, and I am sure they can figure out ways to use almost anything for commercial applications.[75]

Case Study 4: Strategic Computing

Viewed with the idea of development trajectories in mind, from the very beginning VHSIC contained many elements that would be expected to inhibit the commercialization of VHSIC technology as well as the capacity of VHSIC participants to benefit from any advances made in their own separate commercial operations. Certain military end-use requirements for VHSIC components diverged significantly from the established needs of large commercial users. The Pentagon created a dedicated network of VHSIC suppliers and outfitted them with specialized production lines. Military policy also truncated the extent to which VHSIC technology could diffuse to commercial users outside of the program. And, again, VHSIC occurred in the 1980s, when a set of commercial development trajectories for semiconductor technology had already been well established in the United States, Western Europe, and Japan.

Similarly, the Strategic Computing Program was constructed from the very beginning to emphasize specific military applications rather than generic research. Like VHSIC, the SCP's attention to military needs on the demand side created a military-specific development trajectory that was soon entrenched, on the supply side. As the program's emphasis on specific military end uses shifted the program's main focus from research to development, the bulk of the program's work shifted from university labs to large military contractors. Even in the absence of widespread classification, this shift ended up confining the diffusion of information about technical advances to a narrow circle of military-oriented firms.

Like VHSIC, DARPA's SCP was based on the Pentagon's growing realization that military and civilian technology had reversed roles since the 1950s. Technological advances in the commercial sector had far outpaced military efforts to develop several new computer technologies that were each expected to play an essential role in the next generation of high-tech weaponry. Promoters of the program argued that SCP would spur the development of basic technologies that would ultimately find widespread application in both the military and civilian sectors.

On its surface, the program was designed to develop generic knowledge-based (or "artificial intelligence") software and related data-processing approaches. These advances purportedly would allow computers to reason like human experts, to understand human speech, and to recognize objects with machine vision.[76] Proposed in 1983, SCP was set to last ten years, with a budget approaching $100 million per year. At the time, however, the move was widely viewed as the U.S. government's primary response to Japan's ambitious Fifth Generation computer project and, as such, it quickly became mired in partisan bickering over the federal government's appropriate role in the economy.[77] In order to sell the program to Congress and to others within the Reagan Pentagon, DARPA chose purely military goals to drive further development of the technology.[78]

DARPA planned to demonstrate the utility of the generic technologies it sponsored by having them designed at the outset into three prototype military systems—an autonomous land vehicle for the Army, a battle-management system for the Navy, and a cockpit computer for the Air Force that could "converse" with pilots and offer advice in the heat of combat. Each of these prototypes had one basic characteristic in common: Each was projected to work by processing knowledge—by "thinking" for itself—instead of merely processing raw numeric data. Each system was thus supposed to drive ongoing development in several areas: knowledge-based software, expert systems, machine vision (for the land vehicle), speech recognition (for the cockpit computer), natural language processing (for the cockpit computer), and parallel processing architectures.[79]

Neither the ambitious technical goals nor the timetable on which they were to be reached was considered realistic by many in industry and academia.[80] Commercial proponents of high-speed supercomputers were especially concerned about the program's heavy emphasis on artificial intelligence, a consequence of the program's ambitious prototype goals.[81] Critics also contended that the achievement of DARPA's goals would depend on "scheduled" technological breakthroughs, even though breakthroughs are by their nature unpredictable. They complained, moreover, that the emphasis on meeting technical milestones for flashy short-term demonstrations would divert money and attention from basic conceptual research.[82]

SCP's computer architecture program resembled VHSIC's program for the development of superspeed chips in that it funded several simultaneous projects in the same area. The system was set up to spur creative competition between research groups and to provide some insurance against the possible failure of individual approaches; it allowed DARPA to play the role of venture capitalist, picking winners and losers from among the best in the field. But SCP had created a set of programmatic goals that effectively isolated the production infrastructure from the pursuit of commercial end-use alternatives. Each competition's preset focus was on developing technologies designed to work in particular military systems.

Observers of the Pentagon's growing role in funding basic and applied research into advanced computing worried, also, that commercially viable spin-offs would be subject to classification and thus restricted to military use. DARPA officials contended that the bulk of SCI's *generic* research would not be classified, since most of the generic research would be done in universities. Nevertheless, advanced computer architectures were to be developed jointly by universities and private firms, and applied product development would mostly take place in private industry. As the program's emphasis on specific military end uses shifted the program's main focus from research to development, the bulk of the program's work shifted from university labs to large military contractors.[83] Even in the absence of widespread classification, this shift ended up confining the diffusion of information about technical advances to a narrow circle of military-oriented firms. Commercially relevant research results remained trapped within companies that lacked commercial divisions that might be capable of nurturing civilian spin-offs.[84]

After six years, opinions varied widely about whether there were many civilian spin-offs to be had. The program helped to develop some basic computer technology, though it was by no means clear that the advances depended on sponsorship by SCP. As for applications, the SCP further funded the development of the Connection Machine, a high-speed computer that uses parallel processing and that was finding commercial applications. But the Connection Machine had been under development initially for commercial purposes, anyway; so in this case SCP appears simply to have played the role of venture capitalist. Otherwise, the program focused on purely military applications, and the results of those were mixed. In 1989, DARPA reduced the program's emphasis on knowledge-based software and discontinued the main autonomous vehicle work at Martin Marietta.[85] Work on a cockpit computer continued and work on a naval battle management system had resulted in a computer that could calculate how best to redeploy ships to compensate for the loss of a part of the fleet.

Spin-Off in Historical Perspective: Lessons from the Case Studies

We have looked at four cases of U.S. military involvement in technological development, two from the 1950s and two from the 1980s. One essential characteristic differentiates the two technologies developed in the 1950s from those developed in the 1980s—in both of the 1950s cases, when military development of each technology was just beginning, neither had a well-established commercial development trajectory or a widely recognized commercial production infrastructure. In stark contrast, the Pentagon's 1980s efforts were motivated by the perception that separate, well-established development trajectories in the commercial sector had already yielded technical performance capabilities well beyond those available in the military sector. Military projects sought to upgrade military technology to commercial capabilities and to yoke further commercial advances to the specific needs of military users. Moreover, the 1980s projects proceeded in the context of a perceived threat from foreign producers to continued American dominance of the technologies involved.

The outcomes of the two 1950s cases differ from one another because in one case military involvement created a military-dominated technology development trajectory and in the other it did not. In the case of transistors and integrated circuits, military end uses promoted a form of the technology that initially converged with the needs of commercial users; this enabled the creation of a commercial development trajectory that evolved alongside the military trajectory, sometimes overlapping, sometimes not. At the same time, military procurement fostered the development of a uniquely independent set of suppliers whose very survival was linked to the constant exploration and expansion of possible (military and nonmilitary) end uses for the new technology. Military policies also actively promoted the diffusion of basic technological information to potential users and suppliers that were not a part of any military program. Thus, when the end-use requirements of military users began to diverge from the mainstream needs of commercial users, an alternative, self-perpetuating commercial trajectory was already well in place in the industry, due in large part to military policies that had prevented the entrenchment of a military-specific development path.

By contrast, in the case of numerical controls, military end uses promoted a form of the technology that diverged at the start from the needs of most commercial users. This military-specific development trajectory was then entrenched in the core of the industry as the Air Force set about creating a specialized network of users and suppliers. It was further entrenched by military policies that inhibited diffusion of the technology to potential users and suppliers that were not a part of the military network. While the large American machine tool makers focused on meeting the needs of large-scale, technologically advanced military contractors, their competitors in Japan were developing close links to large industrial firms and their networks of small and medium-sized subcontractors. When the development of the microprocessor and simpler programming languages made the technology of NC widely accessible for everyday machining operations, the Japanese industry was better configured than the American industry to mass-produce low-cost NC tools.

The outcomes of the two 1980s cases differ from the outcomes of the 1950s cases because both projects sought to develop the latest generation of a technology that was already under development (both here and abroad) in the commercial sector. Because of that, both programs naturally focused on military-specific forms of the technology that had no commercial markets and were therefore not being developed by commercial firms. In both cases, military-specific objectives came to dominate the objective of dual use. In the case of VHSIC, the priority of immediate systems applications shifted the program's focus from the semiconductor merchants to the military system houses; a similar shift from university labs to military systems houses occurred during the SCP as generic technologies were tailored to the needs of specific military products. In the end, both projects were moderately successful at developing the applications that were desired by the military services and basically unsuccessful at creating dual-use technologies or commercial spin-offs.

Conclusion: From Spin-Off to Spin-On

The issue is not whether spin-off will continue to occur or, indeed, whether spin-on and spin-off can coexist; of course it will and of course they can. In terms of long-term competitiveness, the critical issue has to do with which process is faster. Which provides the quicker route toward full exploitation of the technological complementarities between military and commercial applications?

The United States can no longer afford to divide its technology base into two separate entities, one for the battlefield and one for the marketplace. Despite their demonstrated capacity for saving American lives, most of the high-tech weapons used to such stunning effect in the Persian Gulf war were based on technologies more than ten years out of date. In the past decade, technological leadership in specific areas has passed to others. When allied commanders urgently needed spare battery packs to power their command and control computers in the midst of battle, they had to send to Paris and Tokyo to get them.

For the foreseeable future, America's military technology base will be shaped by three trends: declining military budgets, an expanding overlap of military and commercial technologies, and an increasingly global marketplace for high technology in which the much larger and more dynamic commercial component determines the direction of innovation. There will be more spin-ons from commercial producers to the military sector and fewer spin-offs in the opposite direction.

These trends would matter less for America's security position, as Moran and others have argued, if military-relevant technologies were in fact readily available for purchase from a sufficiently diverse set of international suppliers.[86] As Chapter 1 argued, however, market forces already are allocating to other countries entire industries (or segments of industries) that are crucial for producing the next generation of sophisticated weaponry. Given existing accumulations of market power and technological experience, a small number of firms in Europe and (especially) Japan increasingly possess the capacity to dictate the degree of access U.S. producers have to essential technologies with substantial spin-on potential.

But the case for maintaining a domestic supply base goes beyond familiar, and in our view well-justified, concerns over access for U.S. producers to technologies that are actually traded in the international marketplace. There is concern, as well, that domestic firms gain access to the type of technical know-how that is not so easily traded, either through markets or scientific conferences. Today's most advanced commercial suppliers of high-volume, high-tech products gain this sort of know-how by actually introducing products and then fine-tuning product configurations and volumes to actual demand. Due to its evolutionary and partly tacit nature, much of this knowledge cannot be embodied in a product or a piece of production equipment. It cannot be described in a blueprint. It simply accumulates in the firm or network of firms where it originates, creating a kind of localized knowledge base, a unique and relatively protected source of comparative advantage.

The argument for pursuing a spin-on strategy is, at its root, an argument in favor of maximizing the available opportunities for generating and diffusing this kind of technological *learning* throughout the American economy. When military contractors remain isolated from the product performance and low-cost manufacturing demands of high-volume consumer markets, or when the Pentagon attempts to harness the cutting-edge expertise of commercial firms only to yoke their development efforts to the creation of low-volume, military-specific product applications, the DOD sacrifices the opportunity to benefit from the important technological spillovers that now run from commercial to military products. Defense firms will benefit especially from commercial expertise in the organization of production for small- and medium-sized lot batch manufacturing—especially appropriate for making the components, subassemblies, and subsystems that account for a considerable portion of the value added in military systems. Civilian suppliers will benefit too, by having a demanding, sophisticated defense-sector customer—as long as they are not required to design any military-specific capabilities into their products (and assuming burdensome and unnecessary DOD procurement regulations have been removed).

It is not enough to focus Pentagon programs on "stage 1" (research and feasibility) areas with high potential for military payoff, but few commercial prospects. The United States can no longer meet its need for state-of-the-art military technologies without competitive industries in the relevant commercial sectors. As noted in Chapter 1, the latest generation of consumer products—camcorders, electronic still cameras, compact disc players, and hand-held TVs—plus new high-volume products developed for office and home use—portable faxes, copiers and printers, electronic datebooks, laptop computers, optical disk storage systems, smartcards, and portable telephones—have much in common technologically with engineering workstations, telecommunications networks, military avionics, and other gadgetry of obvious and increasing value to the U.S. Department of Defense.

Most important, the development of these products for mass consumer markets creates a huge demand for low-cost, high-quality production of components and subsystems, most of which are critical for both military and industrial uses. These include everything from semiconductors and storage devices to packaging, optics, and interfaces. For instance, the new high-volume products contain a wealth of silicon chip technology, ranging from memory and microprocessors to charge-coupled devices. In addition, commercial firms already produce enormous quantities of sophisticated optoelectronic components—including laser diodes and detectors, and LCD shutters, scanners, and filters—for use in mass market applications such as compact disks. Japanese miniature TVs are the leading-edge users of flat-panel, amorphous silicon-thin film, liquid crystal display (LCD) technology, and other interface technologies that are soon to be widely applied in both industrial and military systems.

Consumer applications also drive the development of militarily significant storage and packaging technologies. Advanced digital audio tape is as dense a storage medium as high-performance computer disk technology; optical storage is beginning to spread into industrial and military data applications after first being refined for consumer use in compact and laser disks. High-volume manufacturing requirements have driven the development of innovative packaging technologies that range from tape-automated bonding and chip-on-board to multichip modules; producers of hand-held LCD TVs already use packaging technology as sophisticated as that being used in the most advanced American military systems.

Finally, new consumer products are sparking innovations in precision electromechanical and feromechanical components such as motors, gears, and switch assemblies, as well as recording heads, transformers, and magnets. Other significant examples of sophisticated inputs that are commonly found throughout the consumer electronics industries range from the high-quality lenses used in electronic still cameras and camcorders to the low-end print engines that drive desktop copiers and laser printers.

Economies of scale in the manufacture of such products make it needlessly (and perhaps unaffordably) expensive for military contractors to produce independently of the commercial industrial base. Participation in global high-volume markets now gives private firms the capacity to amortize the development and manufacturing costs of technological inputs that were once thought too expensive and risky for any entity other than the Pentagon to support. Massive demand for new products creates similar advantages for suppliers of components and other technology inputs, too, both because they can drive down per-unit costs through scale economies in production and because they can spread the costs of research and development across a much higher volume of sales. Most importantly, however, spin-on is likely to work faster than spin-off; market forces will diffuse technological advances more quickly than the military bureaucracy.

In short, the continued bifurcation of the U.S. technology base creates serious security risks that only intensify as high-volume consumer products become technologically more sophisticated. Military product markets are typically too small, and the pace of product introduction too slow, to generate the

same breadth of cost-reducing, knowledge-enhancing manufacturing experience that accrues, in the same amount of time, in the commercial sector.
DOD's task, then, is to gain access to the commercial knowledge base without
compromising it. Americans can conjure many potential threats to their well-
being, but only one technological arsenal with which to meet them.

5

Third World Military Industrialization and the Evolving Security System[1]

KEN CONCA

Introduction

If, as the preceding chapters have argued, the economic and technological foundations of the international security system are indeed shifting, an important question remains: Will the emerging tri-polar economic architecture remain limited to three poles? The past two decades have seen a dramatic upsurge in Third World military-industrial activity; a growing number of advanced weapons systems are being licensed to, produced by, and in a few cases even designed in, the less industrialized countries. What are the future prospects for continued or accelerated military-technological development in the Third World, and how will that development affect the emerging economic architecture of the international security system?

To ask whether such developments will affect the structure of the emerging system is in fact to ask two distinct questions. The first question is whether one or more of the most industrially and technologically advanced Third World countries will develop capabilities sufficient to be an influential force in the emerging system. The second question is whether various developments in the less industrialized countries will, collectively, add up to a significant impact on the system. Such could be the case, for example, if Third World arms production were to grow to the point of undermining the market position of the traditional suppliers, thereby restructuring the international arms trade.

This chapter considers these questions in light of Brazil's recent history of military-industrial expansion. Brazil is often cited as a "leading indicator" of growing military-industrial capabilities in the Third World; thus, its experience may shed light on the collective potential of Third World military industrialization to affect the emerging system. And if any individual countries in the Third World are to emerge as significant economic actors in the security realm, Brazil—the world's ninth largest capitalist economy, one of the most technologically and industrially advanced countries of the South, and a leader among Third World arms exporters—is certainly a leading candidate.

141

In spite of these seeming advantages, there are a number of reasons to doubt Brazil's ability to expand or even sustain the military-industrial growth it has seen in recent years. Foremost among these are the increasingly untenable position of a military-industrial strategy developed under global-market and domestic-political conditions that no longer apply, and the considerable difficulties of adjusting simultaneously to significant changes on both of these levels. Evaluated in the light of future possibilities, Brazil's experience serves as a cautionary tale, both for Brazil and for those who would seek to imitate it.

Military Industrialization in the Third World

Neither of the two possibilities just raised—that individual Third World countries will emerge as significant actors, or that collective developments in the Third World will restructure the emerging system—can be dismissed out of hand. A number of individual military-technological programs of potential significance have emerged in the Third World in recent years, most notably those efforts to develop weapons of mass destruction and potential delivery systems for those weapons. In addition to China and India, which have tested nuclear explosives, and Israel, which is widely acknowledged to have attained a nuclear-weapons capability, perhaps half a dozen others are at or near the threshold of nuclear-weapons capabilities. Whereas most of the concern over nuclear proliferation has focused on the destabilizing potential of incipient nuclear rivalries, these programs also raise the possibility of emerging Third World nuclear suppliers. China, for example, has been implicated as a possible source of raw materials and technology for both the Pakistani and Brazilian nuclear programs, and Brazilian nuclear technicians are known to have provided assistance to Iraq's nuclear program in the early 1980s. More recently, a parallel concern has emerged over chemical and biological weapons, based on U.S. government assertions that some fifteen to twenty nations currently maintain some form of chemical weapons capacity, and that ten or more either possess or are in the process of developing biological weapons.[2]

Although a number of Third World nations possess advanced aircraft capable of delivering nuclear warheads, concern over the proliferation of delivery systems has in recent years focused principally on ballistic missiles. India, Brazil, Argentina, Israel, and South Korea are all in the process of developing advanced aerospace programs with ballistic-missile potential. A number of less advanced countries also have programs, including Egypt, Indonesia, and Pakistan. With the exception of Israeli efforts, which stress military applications, the stated purpose of each of these programs is to develop a civilian space-launch capability.[3] The Missile Technology Control Regime, a nonbinding agreement signed by the seven Western summit nations in April 1987, has had some impact in slowing the pace of these programs.[4] The former Soviet Union, China, and the Third World governments just noted above are not parties to the agreement, however, and the commitment of some of the signatories has at times been called into question (as discussed later in the case of French technology exports to Brazil).

Although most of the attention has gone to weapons of mass destruction and their delivery systems, the collective impact of Third World producers on the international trade in conventional armaments is gradually making itself felt. Based largely on technologies transferred from the industrial countries, Third World arms production expanded rapidly during the 1970s and early 1980s. The Stockholm International Peace Research Institute (SIPRI) estimates that the constant-dollar value of arms production in the Third World more than doubled between 1973 and 1984.[5] The Third World share of international arms exports to Third World buyers increased from an estimated 2.9 percent for the period 1970–74 to an estimated 10.7 percent for the period 1985–89.[6] By the mid-1980s, roughly sixty Third World countries were involved in some form of defense-sector production, with nearly forty involved in production of what SIPRI defines as "major weapons systems."[7]

Both production and exports have been highly concentrated in the more technologically advanced Third World countries. In 1984, the eight largest producers accounted for an estimated 87 percent of the total value of weapons systems produced in the Third World, with the three largest producers (Israel, India, and South Korea) accounting for more than 50 percent of total production.[8] These industries provide an increasing proportion of national defense procurement needs and, in a few cases (notably China, Israel, and Brazil), have generated significant export earnings as well.

The Roots of Military Industrialization

In most Third World countries, military industrialization efforts are linked to both defense and development goals. If recent trends in the industrialized countries led policy makers to question the effectiveness of military-industrial strategies to promote economic growth and technological development, the same cannot be said for the Third World. Although the international arms trade slowed somewhat over the second half of the 1980s, Third World governments continue to spend vast sums on the military (almost as much, in terms of percentage of GNP, as the developed countries).[9] And in many countries, these expenditures increasingly are oriented toward programs of military-technological and military-industrial development. The skepticism of academics in industrialized countries notwithstanding, the belief remains widespread that such programs can provide tangible economic and technological benefits as well as enhanced military security. The issue is not posed as one of guns versus butter, but rather of finding a single path of industrial and technological development that can yield more of each.

The idea that the military-industrial sector can be a springboard for both defense and development needs has been particularly strong among the more technologically advanced countries of the Third World. South Korea's efforts to promote heavy industrialization in the late 1970s were closely linked to military procurement and arms-sector development; Chung-in Moon reported that "defense industrialization in Korea crystalized the government's efforts to combine defense policies with economic policies for overall macroeconomic development."[10] The recent decision to co-produce the F/A-18, as a means of consolidating the Korean aerospace industry, is a continuation of that model.[11] Broadly similar examples can be cited in Taiwan, South Africa, Israel, and Brazil—in spite of the very different security contexts of each of these countries.[12]

Perhaps influenced by these examples, several countries whose defense sectors traditionally have stood apart from the civilian economy have shown signs of moving toward greater integration of military and civilian activities. The industrial enclaves controlled by the military in Argentina, for example, became the target of privatization efforts by civilian governments in the 1980s, with one stated rationale being the need to promote efficiency and boost export-led growth in the defense sector.[13] India, a nation whose arms industry has long been a technological and industrial enclave, has also begun to move in this direction. Raju Thomas reported:

> From the early 1970s on, the Indian approach to defense planning reflects a subtle but significant change in the perspective of the relationship between defense and development. Although the level of defense spending remained at about 3.5 percent of the GNP during this decade, efforts were now made both to minimize the adverse effects of defense spending and to implement it in such a way to produce beneficial effects for the development program.[14]

Thomas suggested that "twenty years of defense planning [subsequent to the 1962 Sino–Indian war] resulted in the 'structural unity' of the defense and civilian sectors of the economy." The principal features of this structural unity are increasing involvement of the state-owned firms controlled by the Ministry of Defence in production for the commercial sector, and a growing contribution of civil-sector firms (both private and public) to defense production.[15]

Another example is provided by China, which, after years of subordinating commercial technology development to military efforts, has in the past decade sought

> to integrate the military and civilian sectors of the economy. In the pre-1980 period, the transfer of resources was a one-way street from civilian to military, with the lion's share of resources going to the nuclear-weapons effort. The goal of the reform program is to establish a two-way flow, focusing initially on transfers from the military back to the civilian.[16]

Thus, although overall defense spending in China has fallen in recent years, military-technological investments have increased.[17]

The trend, moreover, is not limited to the more industrialized countries of the Third World. The Egyptian government, for example, in recent years has sought to promote export-led growth in the defense sector, based on a series of joint ventures and licensing agreements intended to promote assimilation of foreign technology.[18] One of the explicit goals of the defense-modernization program accompanying Indonesia's Third Five-Year Plan (1979–84) was to "provide technological spillover into other areas of industry through the acquisition of dual-use capabilities."[19] And for would-be producers at all levels of current capability, the industrial countries themselves have been powerful examples of governments that perceive a symbiosis between military and commercial applications. The use of Atlas and Titan ICBM boosters to launch Mercury and Gemini spacecraft, for example, was not lost on Third World governments currently seeking to develop both space launch and ballistic-missile capabilities through national space programs.[20]

Although the idea that both defense and development goals could be pursued with a single strategy has played an important role, Third World military industrialization is also a product of the changing structure of the international system itself. To some extent, military industrialization has been an outgrowth of the larger process of economic and technological differentiation that has been occurring within the Third World since the early 1970s.[21] The declining hegemony of traditional arms suppliers (United States, Soviet Union, Great Britain) and the emergence of new competitors (beginning with France in the early 1970s, later joined by Italy, West Germany, and other European suppliers) has produced a restructuring of the global arms economy.[22] The main features of this restructuring have been intensifying competition, an increasingly commercial logic for arms sales and defense-technology transfers, the growing internationalization of production, and the growing importance of technology as a medium of exchange. Multinationals eager to reduce costs, recoup R&D expenditures, penetrate local markets, and establish regional sales platforms were met by Third World governments eager to gain access to advanced technologies. Governments of the industrialized countries, eager to support their defense industries and pursue regional goals, showed little interest in opposing this convergence of interests. The principal recipients of the resulting "offshoring" of arms production were the larger and more industrially advanced Third World countries.[23]

If military industrialization is partly an outgrowth of changing system structure, another essential ingredient has been a set of domestic political conditions broadly conducive to military-industrial expansion. In evaluating how political systems have shaped the possibilities for economic growth in the so-called newly industrializing countries (NICs), Stephen Haggard concludes:

> Though international pressures weigh heavily in developing countries, the ability of such countries to formulate coherent responses is contingent on institutional arrangements and capabilities, among them the nature of economic decision-making structures and the policy instruments available to political elites.[24]

This general observation about the importance of domestic structures is borne out in the specific case of military-industrial growth. Though they are clearly not alone in this regard, authoritarian regimes of the sort that predominated in Latin America and Asia in the 1970s and 1980s often display a natural affinity for militarily oriented development efforts. Perhaps more importantly, they have also proven themselves able and willing to shield fledgling defense sectors from domestic political pressures and economic vagaries, and to steer resources preferentially in directions conducive to defense-sector growth. The potential political benefits of such policies are both symbolic and material. On a symbolic level, military-technological developments provide a strong message of national autonomy; in more pragmatic terms, the emergence of a military-industrial sector can solidify important political bonds among the military, civilian industrialists, and transnational interests. This combination of factors can and does produce a powerful and mutually reinforcing effect: Technology development efforts are pushed to serve military applications, and the military-technological sector becomes institutionalized as the launching pad for broader efforts at technology development.

Thus, although the experience of individual countries has differed, it can be said that the past two decades of military-industrial growth in the Third World have been driven by a convergence of broadly favorable global-market and domestic-political conditions. An evaluation of the viability of military-industrial strategies for the 1990s must therefore start from the recognition that military industrialization occurs at the intersection of global markets and local politics. Can institutions, organizations, and industries established under conditions of market expansion and broad political autonomy adapt simultaneously to new economic and political conditions? Is there a single technological path that will satisfy both the demands of an increasingly competitive, internationalized market context and the demands of the political coalition of domestic interests supporting military-industrial development? It is the availability of such a path that will dictate whether Third World military industrialization will matter in global terms in the 1990s and beyond; thus, it is with these questions in mind that we turn to consider the Brazilian experience.

The Case of Brazil

Brazil possesses one of the largest and most advanced defense sectors in the Third World.[25] Production has expanded rapidly; where two decades ago there was almost total defense-sector import dependence, there exists today a large and technologically sophisticated military-industrial infrastructure, established as a leader among Third World arms exporters. This quantitative expansion has been matched by a growing qualitative sophistication. The simple, inexpensive, and durable systems with which Brazil's international reputation was established in the late 1970s and early 1980s have been succeeded by more technologically ambitious projects. These include the AM-X, a ground-attack aircraft being co-produced with Italian aerospace firms; the Osório, an advanced battle tank; the Navy's submarine program, based on German-licensed technology; and the VLS, an Air Force–managed effort to build a satellite launch vehicle (which would also yield a crude ballistic missile capability).

The Brazilian defense sector consists of three principal industrial segments: the state-controlled aeronautics industry, which produces a range of civilian and military aircraft; the largely private-sector armaments industry, whose principal products include armored vehicles and rocket-launching systems; and the naval construction industry, including both private shipyards and the Navy's Rio de Janeiro arsenal. The principal firms are Embraer, the state-controlled aircraft producer; Engesa, a private but state-supported firm that until recently claimed to be the world's largest producer of wheeled armored fighting vehicles; and Avibrás, a private aerospace firm whose products include rockets, missiles, and fire-control systems. In 1987, Engesa and Embraer joined with Imbel, the state-owned munitions factory, to establish the joint-venture firm Órbita, a high-tech aerospace endeavor concentrating on missile development.[26]

In addition to the aeronautics, armaments, and shipbuilding industries, the military-industrial sector includes ambitious nuclear and space programs that, although not yet fully consolidated in industrial terms, are among the most advanced of their type in the Third World. The space program is divided between the civilian Institute for Space Research (INPE) and the Air Force's Aerospace Technical Center (CTA). Brazil began launching experimental rockets in 1965 and recently inaugurated a new, equatorial launch cite; the Brazilian Complete Space Mission calls for the construction (by INPE) and launching (by CTA) of a Brazilian satellite in the early 1990s.[27] Since its inception, the Brazilian space program has built cooperative scientific and technological links with the United States, several European nations, and, more recently, China. In 1988, Brazil and China signed a space-cooperation accord envisioning a more ambitious satellite-construction program, and in 1989 Avibrás formed a space-launch joint venture with China's Great Wall Industries.[28]

The military's nuclear R&D activities—begun in the 1970s but not officially acknowledged to exist until 1987—actually consist of separate programs for each of the three branches of the armed forces. The Navy is currently developing an enriched-uranium reactor for a nuclear submarine at the Institute for Energy and Nuclear Research (IPEN) in São Paulo. A uranium-enrichment facility, operating outside of the international safeguards system, was inaugurated at the Navy's Aramar research facility in Ipero in 1987.[29] The Army and Air Force also maintain programs of nuclear research, with the former seeking to develop a graphite-moderated reactor and the latter maintaining a laser-enrichment research program at its CTA facility in São José dos Campos. These military-nuclear programs have proceeded during a period of stagnation for Brazil's civilian nuclear power program, which is based principally on a series of technology-transfer and reactor-construction accords signed with West Germany in 1975.[30] Indeed, it was dissatisfaction with the pace of technology transfer and assimilation through the German accords that led to the establishment of the military programs in the late 1970s.

Due in part to the secrecy surrounding these industries and programs, the size of the Brazilian defense sector can only be estimated. Until recently, the conventional wisdom was that the sector involved some 300–350 firms, with total employment of as many as 200,000 workers.[31] More recent estimates suggest that these figures may be significantly overstated, with more likely totals being 100–150 firms (including perhaps 50 dedicated solely to defense) and employment of perhaps 50,000 workers.[32] As official data are not available, the value of arms exports in recent years has also been the subject of dispute. By the mid-1980s, arms sales were widely reported (both in the Brazilian press and in many international defense publications) to have reached the level of $2–3 billion annually. (For comparison, officially reported merchandise exports in 1986 were $22.4 billion, and manufacturing exports to less developed countries, including oil-exporting countries, were $3.9 billion.[33]) These frequently cited figures on arms sales, however, now appear to have been dramatically (and perhaps deliberately) inflated; the most comprehensive independent estimate to date suggests that arms exports peaked at $569 million in 1987 and averaged just over $300 million annually for 1982–88.[34] The principal clients in arms sales have been other Third World governments, principally in the Middle East, but also a number of countries in Africa and Latin America.

Security and Development

The goal of building a modern military-industrial capacity was first articulated by elements within the Brazilian army in the early decades of the twentieth century.[35] Based in part on the desire for an autonomous arms supply, the goal was also an expression of more general developmentalist, modernizing concerns that generated growing unrest within the army in the 1920s and 1930s. These concerns would ultimately grow into a belief, widely shared in military circles, that national security and modernizing development were inseparably coupled. By the late 1930s, during the military-backed dictatorship of Getúlio Vargas, this linking of security and development had become a principal justification for political intervention, and the military came to occupy a prominent role in economic planning and policy formulation.[36] One early articulator of this vision was General Góes Monteiro, military chief of the revolt of 1930 that brought Vargas to power and later army chief of staff and minister of war. He wrote:

> [The Army] is an essentially political organ . . . General policy, economic policy, industrial and agricultural policy, the system of communications, international policy, all the branches of activity, production, and collective existence, including the instruction and education of the people, the political-social regime,—all ultimately affect the military policy of the country.[37]

After World War II, military thinking on the links between security and development took on a decidedly Cold-War flavor, as reflected in the intense doctrinal activity of the Escola Superior da Guerra (ESG) during the 1950s.[38] As with earlier variants, the conceptual definitions of "security" and "development" implicit in the ESG doctrine that emerged during this period were consistent with a particularly expansive notion of the military's role in society. At the same time, influenced by the experience of the Brazilian military's expeditionary force in Italy during World War II, doctrine came to place far greater emphasis on the technological variable as a key component of strategic planning and industrial mobilization. The postwar period also saw the beginnings of a national policy for science and technology, with the military taking the lead role in stimulating research and development in aeronautics, electronics, and nuclear technology.

Global Markets and Local Politics

The military coup of 1964, leading to two decades of direct rule by the armed forces, was a watershed event for Brazil's military-industrial sector. The coup produced a military regime, supported by a largely civilian and increasingly technocratic bureaucracy, with both the intent and the political autonomy to launch the nation on a dramatically different developmental course. One component of that new course would prove to be an enhanced role for militarily strategic industries.

Domestic political conditions in the post-coup period not only facilitated military-industrial expansion but also dictated the nature of the coalition of civilian and military interests that would come to support such expansion. On the one hand, the weak and state-dependent nature of Brazil's entrepreneurial class meant that the incentives and initiatives for military-industrial expansion would have to come from the state itself; on the other hand, the military's dependence on civilian technocratic support preserved an important role for a civilian-oriented, commercial logic in guiding sector development.[39]

Regional political conditions also provided important breathing space for the unfolding of this civil-military approximation and made it possible to establish long-term defense-sector policies. The potential for interstate conflict on the South American continent was minimal during the 1960s, given the spread of inward-looking authoritarian regimes preoccupied with internal security and the continuing U.S. role of regional policeman and balancer.[40] This in turn freed the Brazilian military from a more conventional external-defense function, allowing defense-sector policies to reflect long-term growth considerations rather than short-term supply needs.

Given the prevailing economic crisis when the regime took power in 1964, military-industrial policy in the immediate post-coup period emphasized defense production as a means to occupy idle industrial capacity and provide a Keynesian stimulus.[41] By the late 1960s, however, the favorable convergence of domestic-political and global-market conditions came to define what would become the two key features of Brazilian defense-sector policy: (1) state-led and state-coordinated growth, characterized by consistent support and policy coordination, and (2) assimilation and adaptation of foreign technologies, emphasizing gradual technological learning and absorption over more immediate production requirements.

Key components of the strategy of state-led growth included direct and indirect subsidies, generous tax treatment, state-sponsored R&D, supportive procurement policies, and domestic market protection of related commercial industries.[42] The state-controlled aircraft firm Embraer illustrates the use of these policy tools. At its inception in 1969, 82 percent of Embraer stock was held by the Brazilian government. Private capital, which showed little enthusiasm initially, was attracted through a tax-subsidy scheme allowing Brazilian corporations to deduct 1 percent of their income tax to purchase Embraer stock.[43] Although the government share of common stock has fluctuated greatly in the more than twenty years of the firm's existence, the government has always retained a majority of voting shares. Virtually all of Embraer's management team and technical staff were drawn from the Air Force's Aerospace Technical Center, ensuring the continuation of a close working relationship between the two.[44] The firm was also helped by tariff protection of the large domestic general-aviation market and Air Force procurement of succeeding generations of military and dual-use aircraft.[45]

The second defining feature of military-industrial expansion during this period was assimilation of foreign technologies. Following the ascendancy of nationalistic hardliners to power within the military regime in the late 1960s, the prior emphasis on the defense sector as a short-term Keynesian stimulus gave way to the vision of a military-industrial engine of longer-term technological development. So-called strategic sectors such as nuclear and aerospace were seen as a principal means of acquiring and assimilating foreign know-how as a step toward greater technological autonomy. This shift was facilitated by the fact that it coincided with the emergence of European defense-technology suppliers on the international scene, lessening the control of the United States and Soviet Union in matters of technology transfer.

The strategy of technology acquisition was carried out through a series of licensing and co-production agreements beginning in the early 1970s. A 1987 estimate suggests that roughly half of all Brazilian arms production can be traced directly to technological inputs from prior licensing and joint-venture agreements.[46] The definitive characteristic of Brazilian licensing and co-production efforts has been insistence on technology transfer, exploiting increasing supplier competition and using domestic-market access as a source of leverage.[47] Important agreements during this period included co-production with Northrop for the F-5 purchase made by the Brazilian Air Force in 1973 as well as licensing agreements to produce Italian Macchi MB-326 jet trainers (1970), British Niteroi-class anti-submarine frigates (1972), West German COBRA anti-tank missiles (1973), and Piper light aircraft (1974).

Changing conditions during the 1970s added a third component to the strategy of state-led growth and technology assimilation. Domestic procurement, though wielded as an effective tool at key moments, proved inadequate to sustain more sophisticated weapons-system production. This forced the sector to turn to export-led growth as the vehicle for its continued expansion. At the same time, oil price increases generated balance-of-payments and budgetary crises that forced the regime to seek new policy instruments in the international realm. The arms and aeronautics industries met these needs on several dimensions; they represented a potential source of foreign exchange, a diplomatic instrument to woo oil-supplying states in the Middle East and Africa, and a vehicle for stimulating multinational participation in technology development efforts.

Growing incentives to promote export-led growth in the arms sector converged with new structural opportunities to do so. The growing commercialization of the global arms market increased competitive pressures on traditional suppliers, further increasing their willingness to transfer relatively advanced technology. And the petrodollar-driven surge in Third World demand created a market niche for relatively simple, durable, and inexpensive weapons systems at the medium-tech level. By the late 1970s, the need to export had become both a fact of life for the leading Brazilian firms and part of an explicit strategy for sector development, carried out through aggressive marketing, high-level policy coordination, strong support services, and cultivation of a reputation as a reliable supplier motivated by commercial rather than political considerations.

This new international context coincided with changing domestic conditions of equal significance. Recession brought an end to an era of rapid growth in the late 1960s and early 1970s (a period sometimes referred to as the "Brazilian miracle"). And by the mid-1970s, a fundamental change was occurring in the tenuous balance between the military and civilian elements of the elite coalition supporting military rule. The political bargain on which the regime was based had from the start defined entrepreneurial capital as junior partner to the dominant alliance between the military and state capital. Growing dissatisfaction with this arrangement on the part of domestic industrial interests led to a growing challenge to state control of the economy.[48] During this period, the defense sector increasingly took on the character of what has been called a "public-private partnership," in which strategic guidance by the state coexisted with substantial private-sector autonomy at the level of business decision making.[49]

As a result of these changes, the defense sector had by the late 1970s entered a new phase, defined by rapid export-led growth and an increasingly commercial rationale for decisions and policies. Strategic considerations of the armed forces continued to shape policy at the highest levels, however, particularly in terms of technological priorities. And potential contradictions between commercial and military goals were largely avoided, principally because of the broadly shared belief within the regime that its strategic priority of technology development was best served through commercially oriented expansion.[50] Expanding market opportunities abroad and a new round of domestic procurement initiatives also eased any such tensions, propelling the sector's rapid growth and capacity expansion in the early and mid-1980s.

In summary, the period following the 1964 coup saw a growing convergence between the military regime's defense-sector policy goals and the structural conditions enabling effective pursuit of those goals. The result was an expanding defense sector organized around state-led growth and the pragmatic but consistent pursuit of technological autonomy. Although changing conditions in the 1970s caused some adjustments to the model, expanding market opportunities allowed these basic organizing principles to remain intact even as the sector came to be characterized by an increasingly internationalized, commercial logic.

The Emerging Tensions of Military Industrialization

The rapid expansion of Brazil's defense sector has been a powerful example to other would-be industrializers. One reason for this is the powerful appeal of the idea that the defense sector can serve as an engine of technology development through the pursuit of strategic priorities that further both military and economic ends. More recently, however, Brazil's military-industrial sector has entered a period of substantial difficulties and growing tensions. Large government budget deficits have cut into funding for military R&D and domestic procurement by the armed forces, and Brazil's massive foreign debt has made it increasingly difficult to finance military-industrial projects through external sources. The combination of financing problems, reduced domestic procurement, and a shrinking Third World export market has produced a financial crisis for the leading defense-sector firms. By early 1990, the combined debt of Engesa, Avibrás, and Embraer had reached an estimated $1 billion.[51] Avibrás filed for relief under Brazil's bankruptcy laws in January 1990, and Engesa, unable to recoup its heavy R&D investments in the Osório tank program, soon followed suit.[52] Embraer, though in a stronger long-term position thanks to the strength of its civilian product line, remains plagued with cost overruns in the AM-X program.[53] Late in 1990, the firm laid off almost one-third of its labor force, announced a severe cost-cutting plan, and postponed indefinitely its most important new civilian project (the EMB-145, a forty-five-seat jet targeted at the growing regional commuter airline market).[54]

Underlying these current hardships is a deeper dilemma: Brazil's balancing act between the strategic and commercial concerns of defense-sector production is becoming increasingly difficult to maintain. Previously convergent global-market and domestic-political forces are increasingly pushing and pulling in different directions, in terms of the influence they exert on the sector and, as a result, on the process of technology development within the sector. Moreover, the resulting tensions are more than just a new contextual challenge to which the sector must adapt; they represent a challenge to the most basic premise of military industrialization as practiced in Brazil.

One important set of pressures on the defense sector is rooted in the changing Brazilian political situation. The role of the military in national political and economic life is changing, as part of Brazil's long and complex transition from military to civilian rule.[55] During the transition process the military has sought to define a new role for itself within the emerging social and political context, while preserving wherever possible the elements of influence it maintained under the old system.

The defense sector is caught in the middle of this political transition in several ways. As the military has relinquished direct political power, a new set of institutional interests regarding technology development have begun to emerge within the armed forces. Though the process has been halting and uneven, military doctrine has begun to emphasize a more externally oriented perspective on defense and security matters, particularly in the Air Force and Navy.[56] More significantly, the defense sector promises to be an important instrument for the military to maintain its political influence in a number of areas beyond defense policy—including foreign trade, diplomacy, macroeconomic and budgetary policy, and science and technology policy. For this reason, "back to the barracks" has thus far not meant "out of the factories"; rather, retaining control of the defense sector has been one of the military's priority political goals in the postauthoritarian environment.[57]

Emerging simultaneously with these domestic pressures have been the increasingly rigid constraints of the global-market context of defense production. The processes of rapid technological innovation, internationalization, and commercialization that had come to characterize the global arms economy by the late 1970s continued throughout the 1980s. But whereas these trends emerged in a rapidly expanding market, they endure today in a context of increasingly intense market competition, characterized by a growing number of suppliers and weak demand.[58] A niche still exists for the more technologically advanced Third World producers, but exploiting that niche requires responsiveness to the rigorous demands of the market and constant attention to competitive position. Given prevailing market conditions, this implies strategies for product design and technology innovation that are driven by market preferences; it also implies a growing reliance on external sources for capital, technology, and markets.

The net effect of these roughly simultaneous domestic-political and global-market changes is to exert a series of increasingly contradictory pushes and pulls on the defense sector. Emerging conflicts revolve around (1) a less certain technological path to be pursued, (2) a renewed debate about the meaning and feasibility of technological autonomy, (3) a shrinking resource base at a time of proliferating demands, and (4) an emerging struggle for control of military-industrial and military-technological decision making.

Filling a Niche Versus Filling a Need

One of the principal emergent tensions involves the path of technological development to be pursued. The gap between what defense planners think they need and what defense firms think will sell in the international market has been growing, and the trend is likely to continue. As technological capabilities expand, these two priorities increasingly suggest different paths of technological development for a number of important weapons systems.

Each of the three services adopted a different response to this problem in the 1980s. The Army's *indústria bélica* came to be dominated by the firm Engesa, which developed a totally export-oriented product line (to the point that Engesa's crowning technological achievement, the Osório tank, is too heavy for Brazilian bridges and of questionable value at best in the context of virtually any imaginable military task facing the Brazilian army). The Navy's response has been diametrically opposed—market dictates have essentially been ignored, thoughts of export-led growth for the most part abandoned, and production concentrated in the Navy's Rio de Janeiro Naval Arsenal at the expense of the private-sector shipyards. The Air Force, aided by the strong performance of Embraer's civilian product line in the international market, has until now been more successful at straddling domestic procurement and export market requirements. The AM-X program, begun in 1980 and beginning to deliver planes at the end of 1989, was meant to fill both a perceived service need (for a ground-strike aircraft) and a perceived market niche (between the more costly U.S. F-16 and the less capable British Hawk Hunter). But the program's spiraling cost has made even breakeven export sales unlikely, and the Air Force and Embraer face a series of stark choices about future military projects. The Air Force has long supported the idea of eventually producing a supersonic fighter—the AM-X program itself was intended to be a step in that direction. In market terms, however, the firm's best option for military production is a more advanced version of the Tucano, a military trainer in production since the early 1980s. Given the firm's current difficulties, both fighter and trainer programs are stalled on the Embraer drawing board; in the longer run they represent mutually exclusive paths for the firm's future military production activities.[59]

The tension between filling a niche and filling a need is also apparent in the growing pressure for domestic military procurement to conform to export-driven product design. Historically, domestic procurement by the Brazilian armed forces has been crucial to the export opportunities of new product lines; procurement has helped initiate production and has provided demonstration value. Embraer in particular has benefited from Air Force procurement of each of its successful export products. The current budget squeeze on military procurement, however, has made the traditional export-promoting role of domestic procurement less and less feasible. The Army's inability to purchase Osório tanks, for example, has so hampered Engesa's efforts to market the tank abroad that the firm offered the Army highly concessional rates and leasing alternatives.[60] Avibrás has faced even greater difficulties; the end of the Iran–Iraq war forced the firm to reorganize under Brazilian bankruptcy law, at a time when scarce resources and changing budgetary priorities have made it impossible for the military to provide support.[61] Even Embraer faces difficulties in this regard, as reflected in the reduction of the Air Force's order for AM-X aircraft.[62] If the military proves less successful in claiming budget resources in the postauthoritarian environment, the domestic appetite for export-oriented products seems likely to decrease even further. Alternatively, if the military can lay a greater claim to budget resources (perhaps as a *quid pro quo* for nonintervention in domestic politics), it seems likely to steer technology development efforts more in the direction of evolving defense doctrine and long-delayed re-equipment desires, provoking conflict with the export sector.[63]

The Ambiguity of Technological Autonomy

The emergence of divergent technological paths has been paralleled by a renewed debate on the meaning and feasibility of technological autonomy. Under military rule, the guiding principle of defense-sector policy was to accept an externally oriented strategy of technology assimilation as a transitional step in the quest for an autonomous, self-generating capacity for military-technological innovation. What has resulted, however, is a sort of treadmill effect: The domestic content of a particular product line increases over time, but high levels of external dependence for technology and components endures across generations.[64]

The problem of technological autonomy is exacerbated by market pressures. Maintaining an internationally competitive (and thus commercially viable) defense sector in the 1990s, particularly in the debt-constrained Brazilian context, will require increasing integration into the global arms economy at all levels. This means continuing and perhaps enhancing the already extensive reliance on external sources for technology, components, investment capital, and markets, and it may even mean offshoring production to less advanced Third World countries (a trend foreshadowed by recent agreements on technological cooperation with Argentina, China, Iraq, Egypt, and others).[65]

The continuation of external dependencies for succeeding product generations, and their entrenchment at all stages of the production process, however, are increasingly inconsistent with the long-standing goal of a technologically autonomous defense sector. By the mid-1980s, firms such as Engesa, Embraer, and Avibrás had in effect linked the entire spectrum of production—investment, design, manufacture, sales, and financing—to the export market. Embraer, for example, imported an estimated $247 million in parts, components, and equipment in order to export an estimated $385 million in aircraft and parts in 1988.[66] The Astros II, Avibrás's highly successful saturation rocket launcher, represents an extreme example of this phenomenon in that it was partially financed by Iraq and designed with the specific needs of the Iraqi military in mind.

At first glance, the Navy's shipbuilding and nuclear programs, which have neither emphasized export opportunities nor evolved in response to external demand, would seem to offer an alternative to this export-linked technological treadmill. In the case of shipbuilding, however, the limited size of the domestic market has thwarted efforts to increase the domestic content of production.[67] And even the Navy's nuclear program, which has been cited as a leading example of a successful path toward technological autonomy, appears to have relied heavily on the diversion of civilian personnel trained as part of the nuclear accords signed with West Germany in the mid-1970s.[68] As with the other services, the Navy has found it increasingly difficult to maintain control and stimulate innovation simultaneously.

Diverging Interests and Proliferating Demands

Diverging technological paths and a renewed debate on autonomy are symptomatic of growing interest conflicts within the military-industrial sphere. As suggested previously, tensions between military and commercial interests were largely avoided in the late 1970s and early 1980s; the military's support for commercially oriented defense-sector expansion, and the expanding international market context, made it possible to satisfy both commercial and strategic interests simultaneously. Indeed, commercial interests *were* strategic under those circumstances. The 1990s, however, present a very different context: The logic of the market generates strong pressures for a privately controlled and commercially oriented defense sector, at the same time that defense-sector control has emerged as an important vehicle for the military in its effort to retain political influence in postauthoritarian Brazil.

Export policy provides one example of emerging interest conflicts within the sector. Closer technological cooperation with the United States retains strong support in some military and civilian circles, and yet Brazil's military-technological policies have been a source of growing tension between the two countries.[69] The United States has sought to make Brazilian restraint in arms transfers, for example, a condition of closer technological cooperation. Brazilian acceptance of such conditions would not only risk losing customers in specific instances, but also threaten Brazil's image as a nonpolitical supplier, considered crucial to export success. The temporary solution adopted when tensions flared over arms sales to Libya in the mid-1980s—restricting only "terrorist-usable" sales (mostly small arms)—did not resolve the problem.[70]

The Iraqi invasion of Kuwait in August 1990 provided the strongest evidence to date of such emerging interest conflicts. It was revealed at that time that a team of more than twenty Brazilian missile engineers and technicians, led by a former director of the Aerospace Technical Center, had been in Iraq for nearly a year, helping the Iraqi armed forces modernize and upgrade their missile capabilities.[71] The incident embarrassed the administration of President Fernando Collor de Mello, which has made closer cooperation with the industrialized nations a central foreign-policy goal. More importantly, it highlighted the fact that unrestrictive arms-sale policies, efforts to develop missile technology, and the military's controversial "autonomous" nuclear program are not only directly contrary to Collor's stated policy goals, but also a fundamental premise of the past twenty years' strategy of military industrialization.

Any divergence of interests within the defense sector is complicated by the emergence of a new pole—that of proliferating civilian demands at a time of shrinking resources. This problem is multifaceted. Brazil's massive internal deficit and external debt have severely constrained resources for military-industrial programs, at the same time that the numerous civilian activities under military control (such as the Air Force's oversight of airports and air traffic control, and the Navy's port-management and coastal-patrol duties) exert substantial demands on this shrinking resource base.[72] Political liberalization has also lessened military influence over budgetary policy, and shifted the relative ability of military and civilian interests to lay claim to scarce resources.

The problem is not just one of budget competition, however; in both the nuclear and space programs, civilian and military research efforts have come into more direct conflict than simply competing for resources. The Brazilian scientific community has led opposition to military-nuclear programs since the late 1970s, arguing that the military's emphasis on nuclear technology development has marginalized needed research in the areas of reactor safety, radiation hazards, and environmental impact.[73] Similarly, the civilian director of the Institute for Space Research was dismissed in early 1989; his proposal to contract a foreign service to launch the first Brazilian-built satellite generated strong opposition within the Air Force, which oversees rocket technology development efforts and controls the effort to build a national launch vehicle.[74] The close links between space-launch and ballistic-missile technology, and the heavily military orientation of Brazil's program, also complicated efforts by both the University of São Paulo and Embraer to acquire supercomputers from the United States.[75]

The Emerging Struggle for Control

Diverging interests within the defense sector, and between the defense sector and other segments of Brazilian society, ultimately hinge on the question of control. In the short run, military control of defense-sector activities was not weakened by the transition to civilian rule. The nuclear, aerospace, and informatics programs were reorganized prior to and during the prolonged political transition of the 1980s, so as to guarantee continued military influence and shield the sector from encroaching civilian oversight.[76] At the same time, a concerted (and highly effective) lobbying effort was undertaken to shape the postauthoritarian legal and political context (most noticeable during the writing of the new constitution).[77]

Although the military's direct control of specific programs endured, however, the political transition meant an appreciable loss of influence over the larger macroeconomic and budgetary context within which defense-sector activities will thrive or wither. Thus, although the sector's "core"—the key research institutes, firms, and relevant bureaucratic organs of the state— remains under military sway, the "periphery"—private-sector firms supplying inputs, tangential bureaucratic organs of the state, civilian research institutes— is becoming less prone to military influence.

The situation of the firm Embraer reflects the resulting dilemma. The Air Force will undoubtedly continue to exert a strong influence over Embraer production decisions for military aircraft. But it has become increasingly difficult for the military to respond to the firm's main problems, which include a weak capital base, large debt burden, and tendency to become entangled in U.S.–Brazilian trade disputes.[78] Beginning in the late 1980s, Embraer's growing financial difficulties generated tremendous pressure on the Air Force either to increase its level of investment dramatically or to privatize the firm. The idea of privatization has been approved in principal, although Air Force officials talk of retaining some control through contractual obligations for military production or some sort of "golden share" provision.

Whether it is possible to attract foreign investment and maintain the desired control, however, remains to be seen. As Embraer's then-president Ozílio Carlos da Silva stated in 1989, "One cannot seek funds on capital markets without giving some voting power to new stockholders."[79] Clearly, the firm's brightest prospects lie on the civilian side of the market; if long-term financial health is to be reestablished, it will be based not on military production but rather on the firm's expanding line of small (under fifty-seat) commercial aircraft.

A struggle for control of the Brazilian space program has also emerged. After INPE's civilian director was dismissed in 1989 for criticizing the military's role in space policy, a proposal emerged to transfer the agency to the Air Force–controlled Ministry of Aeronautics.[80] Although this transfer was not carried out, and INPE currently resides within the Special Secretariat for Science and Technology, the Air Force did control the appointment of a new director. The military has also retained control of the policy-setting organ for space matters, the Brazilian Space Activities Commission (COBAE), which is subordinated to the Armed Forces General Staff.

Possibilities for the Future

Whether military industrialization in Brazil will thrive or wither will depend in part on the sector's ability to control the flow of resources and sustain, in the postauthoritarian environment, its history of preferential treatment. Given the structural difficulties, if not inherent contradictions, of continued development along the current path, however, future military-industrial expansion would also seem to depend on a fundamental restructuring of the model applied over the past two decades.

One component of such a restructuring might be an attempt to broaden the social and political base of the defense sector. Ironically, while the sector's existence beyond the reach of civil society has meant less "interference" in the form of democratic oversight, it also means a narrower base of social support for sector activities.[81] Whereas the postwar military-industrial complexes emerging in the United States and Europe rested on the multiple pillars of government, industry, university, and labor, the Brazilian experience has seen the marginalization of the latter two groups as a matter of policy. Thus, for example, while labor and industry in the United States are currently forging a united front to minimize cuts in defense spending, the Brazilian labor movement has begun to bring consistent pressure for conversion to civilian production.[82] Bringing organized labor and the civilian research apparatus into the military-industrial fold may be necessary to forge a new coalition in support of defense-sector activities.

Another alternative path that has begun to receive attention involves a shift from export-led growth to civilian-led growth, in which civilian product lines replace defense exports in providing the necessary scale to keep militarily "strategic" firms afloat.[83] This pattern already applies to many of the defense electronics firms and much of the aeronautics industry, and could be further boosted by the Collor government's emphasis on technological competitiveness and industrial modernization (though such emphasis has thus far been more rhetorical than real). As with the current model, however, the question of political control remains both central and problematic. The tendency of the Navy's shipbuilding and submarine programs, for example, to limit private-sector participation, concentrating production instead in the Rio de Janeiro arsenal, seems based in part on mistrust of the private sector's oligopolistic power.[84]

A third possibility, and one that is currently being actively explored, is to solicit foreign investment. As part of the proposed privatization of Embraer, the Aeronautics Ministry announced that it would seek up to 40 percent foreign ownership.[85] Similarly, the Army and the Collor administration's newly created Secretariat for Strategic Issues began preparing Engesa for possible acquisition by a foreign buyer soon after the firm's bankruptcy declaration.[86] Even if foreign ownership proves to be the only alternative in reviving these firms, however, the loss of control and the troubling implications for technological autonomy raise serious questions as to whether foreign investment provides a viable alternative model for the defense sector as a whole.

If it proves impossible to broaden the political and social base of the defense sector, to link production more effectively to civilian industry, or to attract foreign investment, the likely result is a reversion to the "military-industrial enclave" pattern common among Third World defense industries. The military's nuclear program, which began in secret and has minimized external linkages throughout its history, may foreshadow such a trend for the defense sector as a whole.[87]

India and South Korea: Different Contexts, Similar Obstacles

Predicting whether collective developments in the Third World will reshape the emerging economic architecture of the international security system, or whether individual Third World countries will emerge as significant actors, is problematic. One of the principal conclusions of this chapter—that local politics and the historical trajectory of military-industrial development do matter, because they shape future possibilities—argues against just such broad-brush treatment. There is reason to believe, however, that some of the broad structural barriers to sustained military industrialization in Brazil are not unique to that country. In terms of technology, financing, and markets, all would-be military industrializers face the same basic global-market context that has narrowed Brazil's options for the 1990s just as it expanded them in the 1970s and early 1980s. And although domestic political conditions and regional security contexts differ considerably from country to country, the basic political conditions necessary for stable defense-sector expansion may be becoming increasingly difficult to sustain in a number of countries.

Consider as examples India and South Korea, two countries with defense sectors of roughly the same scale and sophistication as that of Brazil, but embedded in very different economic, political, and strategic contexts. Unlike Brazil, India has historically faced credible external threats at the regional level; since independence in 1947 there have been three wars with Pakistan and one with China. And rather than sharing the Brazilian experience of extended periods of military rule, Indian domestic politics has been dominated since independence by the broad coalition of civilian interests represented in the Congress Party. These are important differences, and they are reflected in the Indian military-industrial sector.[88] The perceived need to respond to regional developments—such as the Reagan Administration's decision in 1981 to sell F-16 aircraft to Pakistan—has made it impossible to concentrate resources, Brazilian-style, in programs that do not yield immediate advances in military capability. In spite of India's historical emphasis on technological self-sufficiency in the commercial sector, and in spite of some attempts at defense-sector autarky in the 1960s, the option of promoting indigenous military-industrial development has been seen as too costly and too slow for the prevailing strategic environment.

The different pattern of civil-military relations has also had consequences for the Indian defense sector. Although private and public civilian industry has played a far less important role in India than in Brazil, the Indian military also enjoy considerably less influence and autonomy on military-industrial matters than do their Brazilian counterparts. The political and bureaucratic structures of decision making have been such that debate is dominated instead by senior civilian bureaucrats in the Finance Ministry and Defence Ministry, as well as the Parliament and the Cabinet.[89]

Despite its historical difficulties in this area, India is in some ways positioned effectively to exploit the global arms-market changes sketched above. The historical tendency to vacillate between autarky and "black-box" arms purchases has been supplanted by an increasing emphasis on licensed production and technological absorption (beginning in the late 1970s and accelerated in the 1980s). And India's large internal market is attractive to European defense multinationals, suggesting some leverage in technology-transfer negotiations. In spite of these potential advantages, however, a changing domestic political situation and a redefinition of security policy seem, on balance, to be exacerbating rather than lessening the traditional obstacles to Indian military industrialization.

One important set of changes involves the political definition of military tasks, which has been broadened both externally and internally in recent years.

Raju Thomas described the emergence, beginning in the late 1970s, of a "new political perspective regarding security that sought to achieve a complexity of interrelated goals: defense against external enemies, the maintenance of internal peace and security, and the promotion of economic and social development."[90] Externally, security policy has increasingly emphasized the "extended strategic environment," meaning a de-emphasis of the traditional Pakistani threat and heightened concern over broader developments in the Middle East, Central Asia, and Indian Ocean regions. At the same time, the relative emphasis on internal stability has increased in light of various episodes of unrest, a trend not likely to be reversed by the decline of the Congress Party as a hegemonic force in Indian politics.

This shifting set of security policy concerns has been rooted partly in changing domestic and regional conditions, but also in the reorganization of the political structure of decision making for security affairs. The late-1970s' reorganization of the Cabinet-level and Defence Ministry bodies for defense affairs considerably broadened the role of social, economic, and political considerations in security-policy formulation; at the same time, the voice of the service branches in defense deliberations was reduced, both within the Cabinet and within the Defence Ministry. As Thomas reported,

> The move has had mixed results for the economy and for the military. On the one hand, it has provided defense input in the national planning process and ensured that defense allocations do not undermine the economic and political stability of the nation. On the other hand, the new system has eroded military input into defense planning. . . . In the present setup, the needs of defense tend to become submerged in larger national political, economic, and social issues. Although this may seem unfortunate from the military standpoint, it has been deemed essential for economic planning and stability.[91]

A second important change involves the declining hegemony of the Congress Party in Indian domestic politics. The bureaucratic marginalization of defense interests was to some extent offset by the stability of Congress-Party rule and the resulting network of personalistic channels linking interests in defense, industry, science, and the bureaucracy.[92] The decline of Congress-Party dominance in Indian politics, however, may threaten the stability of the traditional system. If the brief loss of power during 1977–80 to the opposition Janata coalition is an indicator, the security policy consensus in India seems likely to fragment and polarize along a number of dimensions, including the Pakistani threat, regional peace initiatives, the role of the nuclear weapons option in India's defense posture, and the consistency of internal security concerns with individual freedoms and democracy.

Sorting out the net effects of these changes for the military-industrial sector is difficult and requires analysis beyond the scope of this chapter. However, adding the effects noted here to the traditional economic, technological, and political barriers to defense-sector expansion in India raises serious doubts about the prospects for sustained military-industrial growth in the 1990s. It seems likely that the expansion of tasks, fragmentation of consensus, continued marginalization of the military from political decision making, and destabilization of traditional channels of interest representation will undermine any such effort. Selected programs, and in particular the space and nuclear programs, are likely to continue to receive priority treatment and support; but broader expansion seems unlikely (and could be undermined further if progress in the space and nuclear realms were to lessen the perceived needs for a conventional defense posture). Although these problems and prospects differ in some important ways from the set of problems facing the Brazilian defense sector, in both cases a formula for sustained military-industrial expansion—one that speaks at the same time to economic, political, and strategic considerations—remains elusive.

The South Korean defense sector is embedded in an economic and political context that differs considerably from those of either Brazil or India. As was the case for Brazil, South Korea is emerging from an extended period of military rule and is experiencing a protracted political transition, marked by gradual liberalization rather than a clean break with the former regime. And as is the case in India, a credible external threat (in this case, North Korea) has strongly influenced perceived South Korean defense needs. Unlike either the Brazilian or the Indian case, however, South Korea's defense sector has been bolstered by an increasingly advanced industrial economy and by the financial and technological support of an important external patron, the United States.

Military industrialization began in the early 1970s as a component of heavy industrialization of the Korean economy.[93] Boosted by government incentives and U.S. technology, Korean defense industries grew rapidly and were supplying the bulk of domestic military procurement needs by the early 1980s.[94] At that time, domestic market saturation and idle capacity led the defense sector to turn to exports; the U.S. Arms Control and Disarmament Agency estimates that military exports rose from $5 million in 1975 to $400 million in 1982.[95]

Beyond the advantages of an industrial economy, based increasingly on capital- and technology-intensive manufacturing, conditions on the Korean peninsula may also stimulate defense-sector expansion in the 1990s. In spite of some thawing in relations with North Korea, there appears to be a broad consensus among South Korean political and economic elites that the external threat remains substantial, and that an increased share of the resulting defense burden will fall on South Korea itself as the U.S. military presence diminishes.[96] Perhaps more importantly, the defense sector seems well positioned politically. By the mid-1980s, the defense sector was dominated by seven of the ten huge conglomerates that together account for 40 percent of Korean GNP; thus, Chung-in Moon argues that

> The fusion of the defense and commercial heavy industries in the hands of these few big conglomerates may transform the emerging military-industrial complex into a powerful political entity. . . . Big business is gradually gaining political power . . . not only because the ruling regime heavily depends on economic performance of big business for its political legitimacy, but also because recent economic liberalization has given more autonomy to the private sector.[97]

In spite of these seemingly fortuitous conditions, military-industrial expansion in South Korea faces a number of obstacles not unlike those described in the case of Brazil. As in Brazil, the tight control that the military has exerted on the defense sector in South Korea may complicate the prospects for sustained expansion. As summarized in a December 1985 report of the Korea Institute for Defense Analyses and the Rand Corporation:

> Although most Korean defense industry is already privately owned, this ownership is only nominal. Management and utilization are tightly controlled by the military, frequently resulting in low utilization rates, high operating costs, and restricted opportunities for conversion to higher yield civil production.[98]

Such control is the logical product of the way in which military industrialization was initially stimulated. Moon, reporting on the defense industry boom during the second half of the 1970s, indicates that

> The insulation of defense industrialization from competing political claims was a result of a highly centralized decision-making system. While other economic policies were subject to pluralist debates among technocrats, businesspeople, and scholars, the policies related to the defense industry (including heavy industry) were confined to the hands of only three persons during the Force Improvement Plan period (1976–79). . . . All decisions on the defense industry were made by these three, and once decisions were made, they were implemented quickly.

The three were the president's economic secretary for heavy and defense industry, the director of the Agency for Defense Development, and President Park himself.[99]

Given the inability of domestic procurement alone to sustain the sector and the continued dependence on external technology, defense-sector expansion in the 1990s seems likely to demand a global market strategy. The need to adjust to highly demanding market conditions seems likely to clash with the pattern of tight military control, yet the likelihood that such control will ease remains unclear. The emergence of a strong political interest within the military in maintaining or even strengthening defense-sector control as the political transition proceeds would not be surprising.

In addition to possible contradictions between market response and military control, the Korean defense sector faces growing ambiguity over the military-technological relationship with the United States. Korean defense firms benefited enormously from U.S. technological assistance in the 1970s, and the bulk of potential Korean arms exports embody U.S. technology.[100] American veto power over such exports, along with royalty requirements, U.S. resistance to offsets on Korean arms purchases, and broader tensions over the growing potential for competition between the defense sectors of the two countries, have clouded the U.S.–Korean security relationship and emerged as enduring (though not necessarily unmanageable) sources of tension.[101] With the domestic political changes occurring in South Korea, and with the possibility that those changes will shift the balance of strategic and commercial interests within the military-industrial coalition, tensions over issues related to U.S. technological dependence seem likely to increase.

Determining whether military industrialization in South Korea or India will face (or faces already) the specific conflicts and tensions described in the Brazilian case must await a more detailed analysis. In all three cases, however, potentially formidable barriers to sustained defense-sector expansion in the 1990s can be identified. Although the specific causes and conditions vary from case to case, a common theme emerges: The coalition of interests in favor of expanding military-industrial activities must respond to changing global market and domestic political conditions, yet the formula for doing so simultaneously and effectively remains elusive.

The conclusions are clear. Although circumstances differ from country to country, a path that reconciles the economic and political requirements for military-industrial growth remains elusive. As a result, Third World military industrialization is unlikely to yield the type of new military capabilities or broader technological development that would redefine the emerging security system. Instead, the tensions, contradictions, and stark choices of recent Brazilian experience are likely to apply more generally, as Third World governments confront pressures for greater competitiveness abroad and greater pluralism and democracy at home. Moreover, this is likely to be the case both for the more technologically advanced countries of the Third World and for the poorer countries that would follow in their footsteps.

These conclusions in light of this book's analysis should not be surprising. As has been emphasized throughout, it is new sources of economic and commercial strength that are redefining the security system. Military industrialization is at best a strategy of a fading era, and military-industrial programs in the Third World represent an increasing burden for governments that face difficult social, economic, and political problems. Given the enormous scope of such problems, the Third World will almost certainly remain a source of international conflict. But the power to define the security system itself will remain in the hands of the industrialized countries.

III

THE CHANGING GAME

6

The Risk That Mercantilism Will Define the Next Security System

STEVE WEBER AND JOHN ZYSMAN

The economic foundations of the postwar security system have eroded. Europeans have asserted their position as a political protagonist, in trade and finance at a minimum. Japan has become a global industrial and financial power. Together with America's deteriorating competitive position and self-inflicted domestic debt, these changes mean that the balance of American leverage and constraint in finance, trade, and technology has been altered. The new distribution of economic capabilities will define new arrangements of power. This book has not been about marginal change or minor relative repositioning; it has been about how the distribution of power among the major world actors is changing fundamentally. The redistribution creates new possibilities for the international system. When politics catches up with the new capabilities, the emerging system of great power relations will look different from that of the past. The new system may be merely an adjusted version of what we have known for forty years; or it may be radically different.

How do we understand what may emerge, and why? More generally, how does a given structure—a particular distribution of capabilities—shape the evolution of a system of relations among major powers and with the smaller states?[1] Even under the most constraining conditions, international political systems are not defined only by the distribution of power. Rather, the specific character of an international system at all times is historically contingent. The particular preferences, world views, and domestic characteristics of the most powerful actor or actors matter. Even in the late 1940s when the distribution of power was maximally constraining on the possibilities for great power relations, the character of those relations was contingent. The system that emerged reflected the dominant preferences and characteristics of the most powerful actor, the United States.

If the shape of the international system was undefined in this earlier era when the distribution of resources was so heavily skewed toward one state, it is even more wide open now that the distribution is much more balanced. Unlike the period at the end of World War II, the United States can no longer implement its preferences unilaterally.[2] Even as American hegemony declined, the Cold War provided a clear definition of a shared threat that held together the U.S.-dominated western security system. This threat was an important aspect of the bargains that fashioned the global economic order and held that order together.[3] Now, with the radical political changes in Central Europe and the collapse of the Soviet Union, conceptions of what constitutes a security threat will be diverse. Consequently, there will be more than a competition of interests among the powerful countries—there will be a debate about the very conceptions of security within which to frame and argue out specific interests.

With the end of the Cold War and the present deemphasis of military competition among the great powers, it is likely that the evolution of economic relations among them will more powerfully influence the character of the international system. One of two kinds of economic arrangements will emerge from the redistribution of resources and the obsolescence of the U.S.-led postwar system: some form of "managed multilateralism" or a regional rivalry, a twenty-first-century regional version of mercantilism. Each has powerfully different consequences for how the advanced countries will conceive and pursue their broader security interests.

These alternative systems are not equally stable. Managed multilateralism can easily tilt toward regional rivalry. The latter neo-mercantilism has a strong intellectual foundation in theories of technology development and strategic trade that imply a first mover advantage in the competition to capture high-value-added industries; consequently, an economic "cult of the offensive" may emerge as regions compete self-consciously for advantage in global markets.

Avoiding this outcome will not be simple. The new ideas about strategic trade and technology cannot be dismissed. They have a solid analytic and empirical foundation and provide demonstrably powerful guidance for domestic and trade policy. The propogation and potential use of these ideas in national policies creates problems that will have to be resolved mutually to avoid confrontation. But these problems do not fit easily into the existing intellectual and political framework of GATT and will not be addressed successfully there. Though overt hostility among Europe, Japan, and the United States may be hard to imagine, and we certainly do not anticipate that the consequences of mercantilist regionalism would extend to overt hostilities among the blocs, relations could well turn harsh as conceptions of security diverge, security cooperation dwindles, and economic bonds loosen.

Structures of Power and Systems of Politics[4]

Political systems do not change smoothly with changes in the distribution of power. They are regulated by institutions, both formal and informal, that set the context of interaction and establish regular patterns of behavior and expectations. Institutions necessarily reflect the circumstances under which they were formed—but power itself is only one of those circumstances. Even under the most constraining conditions, there are several kinds of order that could provide the foundation for a system of political and economic relations among large powers, and between them and smaller states.

Looking at the case of the American-led system that was set up at the end of World War II throws the point into sharp relief. By stark neo-realist logic, Soviet military power and the accompanying realities of emerging bipolarity should by themselves explain the Pax Americana and the political-economic institutions of the western alliance during the Cold War. It is clear that the United States did face constraints from the need to balance Soviet power, as Roosevelt and Truman discovered to their dismay.[5] But at the same time, there were choices about how precisely to achieve a balance of power. With roughly 50 percent of the world's GNP, an unmatched long-term potential to generate and deploy military power, and a monopoly on the atom bomb, the United States may have had more choices than most other great powers in history have had. East of the Iron Curtain, the character of relations among states reflected Stalin's choice to maximize control over his "allies" and magnify their immediate utility to the Soviet Union in its global struggle with Washington. By the end of 1948, the Soviet Union had fashioned a network of bilateral security treaties that connected Moscow to each of the East European capitals but kept them separate from each other. The organizing principle for this system was "divide and conquer," and its success was to establish each ally's dependency on Stalin for security and economic intercourse. The WTO and the CMEA shared a single purpose: emphasize asymmetrical links between the smaller powers and the hegemon; prevent the formation of a federation or suballiance among the weak. The system did not work quite as anticipated, and it may not have bolstered Soviet power in the long term, but its consequences for the character of economic and political relations among the states that made it up were substantial. The legacy of that system continues to weigh heavily on Eastern Europe and the former Soviet republics even after the collapse of the Soviet Union.[6]

Washington had similar options. The United States could have cut a series of bilateral deals with each of its European allies in 1947 or thereabouts, leading to an American-dominated network of guarantees that effectively would have provided the essential good of security against the Soviet threat. Such a network of bilateral deals organized by and around Washington is not just a theoretical alternative to the NATO bloc that emerged. Robert Taft championed this bilateral approach in the Senate through much of 1947. In March of that year, Great Britain and France signed a bilateral treaty of "alliance and mutual assistance"—the Dunkirk Treaty—which British Foreign Secretary Ernest Bevin later proposed to extend into a network of bilateral agreements that would have included the United States.

Bilateralism would have had several advantages from Washington's perspective. By making security promises on a selective basis, the United States could have maximized its bargaining leverage vis-à-vis each of its potential allies by threatening to exclude individual countries. The United States could then have gained greater concessions in other areas where interests continued to conflict; at a minimum, the United States would not then have been constrained to make equal commitments to unequal allies. It could have claimed a higher price for providing security to relatively exposed and weak states than to strong and well-protected states. Stalin's "extractive bilateralism" was, of course, an extreme version of this approach, and the United States would not have had to go so far.

Simply-put, the organizing principle that characterized NATO—that security is indivisible, and an attack on one is an attack on all—was in no sense determined or even driven by the distribution of power. Security in the West became a nonexcludable good, but only because the United States chose to fashion its relationships with its allies according to that principle.[7] Powerful states can always be expected to set terms for the international system, but not all powerful states are alike.

The principles of indivisibility and nonexclusion that underpinned NATO were even more central to the economic system championed by the United States at the end of World War II. Consider, for example, the "free-trade" model that was institutionalized in GATT. The goal of GATT, albeit compromised and hedged in practice, was to develop an open trading system based on an extension of the "open door" ideal. Discriminatory, preferential trade barriers and currency arrangements would be made transparent and then gradually reduced overall. There were exceptions, most notably for agriculture. There were escape clauses, which allowed states to compromise between open trade and the difficult domestic politics of adjustment. But the MFN (most favored nation) mechanism, by which tariff reductions negotiated bilaterally or otherwise are extended to all other parties to the agreement, became the accepted standard. This was very different from the interwar system of bilateralism in trade and monetary clearing arrangements set up by the Germans to prey on and extract resources from smaller and weaker states, as well as the systems of colonial preference (such as the commonwealth) which the British constructed to provide a protected market for its industries.[8]

Again, the GATT and its attendant principles were in no sense determined by the distribution of power among states. The point is simply this: not all hegemonies are alike. To say that a powerful state or several powerful states will try to construct an international system and set up international institutions to further economic and security interests in world politics begs the most important questions: How do powerful states conceive of their interests, and why do the conceptions they have change?

The U.S.-led system was based on a distinctive set of American postwar ideas about world order.[9] Among those who stress the role of ideas as a crucial addition to power for explaining the nature of the American order, there is broad agreement on three points. First, the American conception of world order was neither entirely coherent in its internal logic nor fully realistic, but it did reflect American economic interests.[10] The conception of a multilateral economy reinforced notions of a multilateral security order which in turn reinforced American economic interests. The United States was the unchallenged industrial leader. The private incentives of U.S. firms and the public incentives of the U.S. economy overall meshed neatly with political ideas about the benefits of the multilateral order.[11] Ideas by themselves might not have sustained American willingness to pay the costs of multileralism, nor instilled confidence in second-tier states that we would in fact carry out our promises. But because these ideas harmonized neatly with our material interests, U.S. commitments to both the security and the economic order were believable. Others had reason to put relatively long-term faith in our promises and to pay serious attention to our demands for cooperation.[12]

Second, the American conception of the postwar international system was powerfully influenced by reactions to the shortcomings and failures of earlier policies, and to our interpretation of the causes of the two world wars. This was true again both of the economic order and of the security order that was linked to it, as well as a deeper conception of social order that extended to the domestic characteristics of states that could be "good citizens" in the new international system. When planning for the postwar order began in Washington, the bilateral discriminatory trading system set up in the interwar years by Nazi Germany quickly became the appointed economic villain. Colonial preferences designed in the early twentieth century to bolster the deteriorating British and French empires were not far behind on the list of evils. In security, the United States worked against a return to a checkerboard system of weak guarantees that were easily overcome by Nazi aggression, in favor of some form of collective security system to take account of the uneven distribution of power.

Third, the United States sought to reconstruct domestic politics in the countries it defeated to promote the development of pluralistic and open societies via representative government.[13] The broad ideas behind the policy packages designed for political reconstruction, imperfect as they may have been, were optimistic and simple, and thus provided a rough guide for action. The linchpin notion was that pluralistic and democratic states that trade freely with each other and share a commitment to collective security can live together peacefully and grow economically without threatening each other's interests or undermining the bargains that were the foundation of co-existence. Peace, prosperity, and democracy might not be shared equally, but they could be shared by all.

These ideas were not accepted wholesale by all states, not even by all states outside the Soviet sphere. But the United States's capabilities were more than adequate to make attractive side payments and occasionally to compel others to accept them, or at least to work within them. The system that emerged did not perfectly match the American conception, but it came remarkably close.

The system was also remarkably stable and robust. It was maintained in its essentials, surprising both academics and others who expected the system to deteriorate sharply with the relative decline of American power in the 1970s. There was still the shared security threat from the Soviet Union, though, a threat which only the United States was in a position to check. The United

States maintained its position as the most powerful international actor and remained willing to provide collective goods even though doing so could no longer be justified on a strictly economic calculus and may have begun to undermine our economic position. Other states helped by beginning to pick up pieces of the burden of sustaining the system. They did so in part because the ideas that formed the basis of the American system gained abroad a legitimate intellectual and political base of their own. By the mid-1980s, academic discourse had mostly turned away from the dangers of post-hegemony disorder and toward new expectations of compromise among a few powerful states that would sustain the central elements of the American-led system beyond the distribution of power that originally had made it possible.

The present security order may not be so easily maintained in the future. Not only has a new distribution of power gradually emerged, but the changing technological profile of production that this book describes may alter the sources of power in ways that could leave the United States short of *primus inter pares* status in not very many years and undermine the economic fundamentals of the postwar order. The intellectual rationale for the multilateral order has also weakened, in part by that same industrial revolution. Finally, the end of the Cold War, the rebirth of Central-Eastern Europe, and the collapse of the Soviet Union itself dilute the security "glue" that might have otherwise held the system together past the point of revolutions in economics and ideas. As the next section suggests, new possibilities are emerging for an international system that fundamentally recasts the character of relations among the major states.

Three Scenarios of the Future[14]

The emerging distribution of industrial and technological resources, what we call here the *international structure*, will not dictate the system of relations among the advanced countries. It appears that there will be three primary regional trading groups, but it is not yet evident what the relationships among them will be. If the industrial and technological base of each region is roughly equal, as we believe it will be, the range of possible arrangements among the blocs is quite broad. The system of relations that emerges will reflect the ideas and values of those who establish the principles and institutions of the new order, but it is not yet obvious who that will be. Those principles, ideas, and values will not reflect the preferences of a single nation. Rather, new notions will emerge from a diffuse set of concerns that reflect different priorities in each region. There often are different problems and priorities for the several countries within each region. Moreover, the institutions of the new system will be built in response to particular crises. The order of crises will also influence the institutions and principles that emerge. In short, the character of the new system is extremely contingent.

In our view, much about the next security order will turn on the character of economic relations. That is, economic relations among the blocs may set the parameters within which security and military issues are dealt with and resolved. Market ties have long been thought to limit conflicts among the advanced countries, and that may continue to be the case. But it is also possible that the frictions of trade and foreign economic policy disputes accumulate into more central enduring conflicts that create serious divergence in the advanced countries' perception of their security needs. Sketches of three model scenarios—managed multilaterism, regional autonomy that is a sort of defensive protectionism, and regional rivalry that is a form of twenty-first-century mercantilism—begin to clarify these alternate economic futures and their implications for security relations.

The most optimistic future is one of managed multilateralism. This amounts to an extension of the post-war American system into a new era in which power is more evenly distributed. Although the United States may no longer be willing or able to pay the costs of sustaining this system, a managing alliance of the United States, Europe, and Japan is possible—with the G-7 as its functional form.[15] Basic assumptions—part fact and part fiction—endure: free trade will generate the expansion of all economies if only each will bear the strains of adjustment to competition; governments will negotiate about the rules of trade, leaving the market to settle the outcomes; and trade will be conducted by private actors in markets according to prices that are set by a free interplay of supply and demand.[16] The central, definable premises of the system remain in place. First, the guiding principle of the trade regime remains the notion that concessions made to one participant would be applied to all participants in a nondiscriminatory fashion—the notion of most favored nation. Second, since the operative assumption is that governments will negotiate only about the *rules* of trade and leave the market to settle the *outcomes*, governments would expect and permit each other to intervene in markets, but only as a last resort and only temporarily as a means of delaying and easing but not circumventing adjustment to international price signals.[17] In some limited number of sectors, adjustment subsidies to firms or communities, controls on imports, or openly managed trade might be politically necessary. However, these cases would be posed as *exceptions* to general rules and would be recognized by all as such. Third, with a general and shared faith that major economies would abide by the rules and share in the benefits, compliance questions could be handled amicably. Sanctions would be rare, since the order would be enforced more by the fear of general collapse than by specific retaliation for presumed violations. In sum, the whole enterprise continues to rest on the conception—really a central assumption—that increasingly open trade would benefit all the advanced countries.[18] Multilateralism would remain embedded in the institutions of the international economy because it would be justified by theories of the gains from trade.

Managed multilateralism on its own logic looks like a reasonable alternative. Certainly the United States can live with this future. The United States adapts some; the Japanese accept some compromises; and the EC makes some difficult choices. It is possible to imagine that all three would benefit absolutely from this arrangement. If the United States adapts well, it remains wealthy and powerful, even if not dominant. If it adapts badly, the risk is lower national incomes and less global influence, but that is all.[19] There might be relative shifts in position among the three centers of power, but the significance of relative gains would continue to be submerged within a set of broadly shared goals and interests in the world economy. The security system that emerges from such an order could be built around collaboration and cooperation among the advanced countries, something approximating a latter-day "Concert of Europe." It would involve some version of joint management by the advanced countries of conflicts in other arenas.

Benign regionalism—defensive protectionism—is a second, though less likely, alternative.[20] The world economy divides into three largely autonomous trading regions with relatively low levels of interdependence. Each bloc would work to limit its exposure to the others, and whatever links were left among them would be managed principally by agreement, not by markets. Markets would separate as trade became more concentrated within each bloc and fell off between them. The result would be three relatively independent economic systems, with low levels of sensitivity to each other's choices and even lower levels of vulnerability to each others' actions. Each region would have its own trading arrangements and financial system. Equally important, each region would have its own supply base, defined in Chapter 1 as a fundamentally indigenous arrangement for the supply of skills, components, subsystems, assembly knowledge, and final product required to ensure independently sustained growth and technological development.

Liberal orthodoxy's prediction that aggregate welfare for the world economy would diminish is probably correct, and benign regionalism may indeed be a world of slower growth as a result of extensive trade protectionism and investment controls that each player employs as it seeks to manage its adjustment. But the more troubling consequences that orthodoxy sees for regionalism—dissension and conflict between the blocs—might not follow. There would be shifts in position and differences in levels of wealth; Japan could end up the richest, with Europe and the United States somewhat less rich—but whatever the precise ordering, all would be wealthy. If the three blocs did not threaten each other's autonomy, the differences between them and the relative gains by any one region might not matter much to the security system. The three internal free zones might well supplant GATT. At best, GATT evolves to become a forum to manage interregional disputes; at worst, it is inappropriate, and there is a void. Three large and inwardly oriented economic blocs could coexist comfortably if mutual desires for regional autonomy, control over domestic welfare, and internal political stability were recognized so that the blocs did not compete for relative power.[21] Friction might develop over the attachment of different peripheral regions to one or another core bloc, but with internal growth and autonomy ensured at home, these disputes would be expected to be marginal. Low interdependence would ensure mostly calm relations.

The third alternative is a regional rivalry that becomes a new twenty-first-century mercantilistic regionalism where the drive for autarky is fueled not by welfare concerns but by worries about relative position and competing state power. The world economy here divides into three largely autonomous trading regions that proceed to compete and to conceive their own security as dependent on a winning position in that competition. They would be intense competitors for a zero-sum prize. The relative shifts in position would not be treated calmly, but would become elements of security concern just as relative gains have been between adversaries in the past. Twenty-first-century mercantilism might not have all the adverse consequences of its earlier manifestation, but security, and the desire to assure autonomy, might reemerge as an issue in the relations among great powers. That development would then bring a different tone with profound consequences for international economics and politics. This would bring a different tone to their dealings. They would be less likely to collaborate and cooperate to maintain order than in managed multilateralism, and less likely to be indifferent to resolutions of security issues within other blocs as we might expect with benign regionalism. The possibilities for conflicts of interest in this world, which we expand upon in a few pages, will at a minimum diminish the potential for cooperative leadership over the world economy and security systems. More likely it would bring new risks of tense competition among the three powers that we have not been accustomed to over the past forty-five years.

Which of these "worlds" may come out of the current juncture? The system that emerges will certainly contain elements of each of these ideal types; but the question is not simply what "proportions" or "features" of each are finally expressed. What matters is which set of principles comes to define relations between the regions. Most likely we face a choice between some version of managed multilaterism or a form of twenty-first-century mercantilism.

Managed multilateralism is based on the presumption that a committee of three can provide leadership for a liberal world previously held together by American hegemony. That, in turn, assumes broad agreement among the three about the purposes and goals of the world economy. Basic differences among these three have been obscured while there has been a clear leader. Agreement about the system's principles, the evidence suggests, is not so deeply rooted and is now challenged in several ways. First, there should be no illusion: American ideas about how to organize a world economy simply are not shared unquestioningly by the European and Japanese. In fact, there are several conceptions of what democratic capitalism or a liberal market economy are and how differences between domestic systems can be reconciled. "France," Joan Pearce wrote in 1980, "adopts the view that in international trade the free market is a chimera and that the best that can be achieved is a balance of interests."[22] Japan operates with different notions of business organization, business–state relations, and domestic international linkages in the world economy. Japanese approaches to business organization favor enduring interfirm network ties over markets as a means of orchestrating interests within a market system, and reserve an important place for the state in creating competitive advantage.

Managed multilateralism cannot easily operate around more state-centric notions of economy. As we shall argue in a moment, the focus will all too easily become national advantage and relative position.[23] At the same time, the diffuse and implicit character of the arrangements that underpin it cannot readily survive a direct focus on trade outcomes—a focus that is increasingly an issue for politics in all three blocs. A privileged group of three—Europe, Japan, and the United States—may in theory be able to provide a collective leadership for the world economy, but not if one of the three at least has a very different conception of what that economy should look like or believes itself to be losing in the market in dramatic ways.

Second, American interests do not continue unequivocally to support free trade, and a shift toward protectionist principles would certainly make it harder to sustain managed multilateralism. In Washington, statements of principle still generally support free trade. Yet from textiles in the 1950s through steel and automobiles in the 1960s and 1970s to semiconductors in the late 1980s, American policy has created such an extensive web of protectionist exceptions that one might ask whether in reality we have a free-trade policy or simply a "reconciliation of interests." Those exceptions came mostly in the form of voluntary restriction agreements in which the exporting country limited its shipments to the American market. Less frequently than formal restrictions, but almost uniformly, the arrangements benefited our competitors and not ourselves. Now even the principles of the multilateral trade are challenged by the United States's adoption in a recent trade bill of the super 301 rules that obligate the president to engage in bilateral trade negotiations for special concessions with those countries thought to be massively disadvantaging American companies. This shift in principle is much more significant than any individual policy or any substantive outcomes that have resulted.

If a common vision of a world economy does not exist to sustain managed multilateralism, and if American commitments to the free-trade system continue to weaken, it is still possible that common aversion or fear of what might replace it could provide the multilateral system support. But two props of the American-led order that might have played that supporting role are at the same time fading. Shared memories of old failures and their lamentable consequences are now generations old. Different conceptions of how to organize world politics may cause disagreement among the major players, but they do not shade over into the aggressive, exclusive, and racist ideologies that accompanied the disputes of the 1930s. Success in economics is no longer so closely tied to the possession of territory and other physical resources that are easily maintained within geographic boundaries, as is argued in Chapter 1.[24] Optimists see these changes as all for the good, and they may very well be, but the dulling of common aversion to traditional risks (by optimism that the future will inherently be more secure) also eliminates an important incentive for continued cooperation. We might also note that the economic system was previously embedded in a parallel security system with the United States at the center of both groupings. Negotiations about one set of questions have always implicitly, and sometimes explicitly, been set against concerns about the other both in Europe and in the Far East.[25] Sometimes over the course of the Cold War, our allies—more single-mindedly focused than we on economic development—were able to exploit our double agenda, and at the same time security concerns bounded the intensity of any economic disputes that emerged. That prop has also been weakened, assuming the Cold War "ends" in Asia to the extent that it already has in Europe. If change itself and the alternatives to managed multilateralism in particular are not to be feared, rival conceptions will, at a minimum, compete more vigorously than they have in the past.

Chapter 1 contended that three trade groups are emerging. If those groups are not connected through intimate ties of trade and investment in a world of managed multilateralism, there is reason to think that they might not stand in calm autonomy, even if they are able to partially insulate their markets from each other. The relative levels of wealth among the advanced countries are shifting; and as Chapter 1 argued, the United States's dominance could give way not to interdependence, but to dependence in domains of technology, industry, and finance. The United States may have to struggle to regain or retain its position and autonomy; those who are newly strong may exercise the control implicit in their market positions. Another possibility is that the United States might make an effort, using policies already in place in Japan as well as Europe, to reestablish its industrial position in ways that could threaten the fabric of interests that hold together the managed multilateral system.[26] The intellectual basis for regional rivalry already exists: twenty-first-century mercantilism that shares with its historical predecessor a competition among nations for industrial and technological position, a competition that shades into an effort by each protagonist to shape the world order to its liking.

Trade, Technology and National Competition[27]

Regional rivalry would be the most dangerous of the possible outcomes for the new international system. Speculating on particular conflicts or scenarios that might bring about one or the other outcome is beyond our scope here. Instead, we are concerned with conflicts that would seem to push the system toward mercantilistic competition, and the forces that underlie them. We examine the major ideas that might come to define specific conflicts along mercantilistic lines by raising the stakes in particular conflicts and redefining interests.

Managed multilateralism will not be sustained by simply repeating the shibboleths of traditional trade theory. The argument that all deviations from the principles of free trade are sinful is under serious intellectual attack. If an adjusted version of managed multilateralism is to survive, it will—at a minimum—have to address and compensate not only for the practical pressures of defensive protectionism, but also for the intellectual pressures that push against open trade and toward a new twenty-first-century mercantilism, especially in high-technology industries. The focus on high technology begins with specific sectors but ends with a concern for the entire economy. The initial focus is on industries that are widely viewed as strategic because they provide inputs that must be available to the rest of the economy. These are the transformative or leading-edge industries that are seen to be changing all other sectors—electronics, biotechnology, and new materials, for example. The development of these transformative industries that act as critical inputs to a wide range of industries depends on relationships with those downstream user industries.[28] The concept of strategic industries then expands to encompass the users of advanced technology, which includes the most vital and productive parts of the economy as whole.[29]

Seemingly domestic concerns about how best to support innovation, sustain the development of science and technology for industrial productivity, and promote advanced technologies become dangerous international issues that could define economic relations among the advanced countries during the 1990s and thus become a defining feature of the international system as well. Indeed, individually, the components of a policy of technological development are already contentious international issues. There are disputes over the reach of intellectual property rights, over dumping practices, and subsidies. When taken together, they raise an international debate about national development, about how to generate and retain advantage in these technologies and industries on which future development and security will rest. Policy choices traditionally considered domestic could become the basis of more serious conflict among the three regions of the modern economy. Domestic discussions about how best to support innovation, sustain development of technology for industrial productivity, and promote advanced technology industries have become a central part of an international debate about national development and, by implication, about which nations will dominate critical technologies and industries in the future.

There already exists an analytic/theoretical foundation for interpreting particular trade frictions in advanced technology industries as fundamental national conflicts. This foundation's conception of the global economy is, surprisingly, a complement rather than an alternative to the liberal economic views embedded in international institutions established under American leadership. The two conceptions really address different aspects of the world economy. But the implications of these two conceptions are radically different. One emphasizes the potential for mutual gain, whereas the other highlights the potential of using state action in global markets to create enduring national advantage.

The intellectual foundations for the notion that trade is really about creating national advantage can be found when we marry the "new" trade theory to theories about technological development that emphasize national trajectories. Here we present the core of the argument that national technology and industry policy are potential weapons in a new form of mercantilistic competition. Our purpose is not to evaluate these arguments as scientific propositions, but to consider their consequences as political conceptions. The "technology trajectory argument" proposes that the current production profile of an economy may powerfully shape possibilities for future development; the "new" trade theory proposes that in oligopolistic markets governments acting strategically can create enduring market advantage for national firms. Each argument has behind it an extensive theoretical and empirical literature widely admired though not universally accepted within the economic profession, and each has important implications for sound domestic policy. Taken together—and increasingly their implications are combined—they suggest that nations can compete for the fast or advantageous roads of industrial development.

As is discussed in the Epilogue, this adds to the agenda of international trade discussions; trade talks must be concerned with how to reduce trade barriers among nations or, increasingly, between regions, but trade talks must also establish principles by which countries can pursue their own domestic technological and industrial development in particular sectors without disadvantaging their trade partners in unacceptable ways. The character of the problem becomes different if one believes that the nature of competition in high-technology industries, or those industries being transformed by so-called high-technology industries, creates possibilities for enduring national advantage. The argument does not suggest that mercantilist trade wars are inevitable, but it does seem that traditional approaches are insufficient. Instead, we believe that the international rules of the road must go further to seek (1) mutual access to national technology institutions, (2) a common component and product supply base, (3) adequate access to national markets for the launch of advanced technology products, and (4) rules about appropriate state support for firms. There are significant obstacles at both the national and international levels to these objectives, and the traditional mechanisms of the GATT may not suffice to overcome them and sustain the system of managed multilateralism.

The Notion of Technology Trajectory

Technology trajectory arguments emphasize the national and local context in which technology develops and in which industrial competition is rooted.[30] They challenge the basic tenets of modernization theory, which depicts a common development course that nations cover at different speeds. According to the old canon, policy could simply accelerate or retard that development. Now it is argued that nations, or regional communities within nations, follow separate development trajectories. Those trajectories rest on differences in industrial structure, social organization, and the role of government in the economy. Different trajectories imply more than that some countries grow faster because of higher savings rates. Rather, industries in one nation, even sets of industries with similar core competences, may make innovations or begin lines of development that create powerful advantages in international industrial competition—advantages that are not readily transferable to others. As a result, there can be decisive winners and losers in international trade; one nation can attain advantages in related sets of industries that its rivals cannot match. Production costs and technology do not automatically converge, but may actually diverge, creating decisive advantage in a range of related industries for those producers rooted in one national environment.[31] This first proposition is, then, that technology evolves along trajectories set by community and market context, correcting the misconception that the direction of technological evolution is driven principally by technical knowledge.[32] Rather, technology develops along a path in which opportunities for tomorrow grow out of research, development, and production undertaken today.[33]

These notions rest on a particular assessment of the nature of technological knowledge and how it differs from scientific knowledge. Scientific knowledge is made up of a set of basic theories, principles, and premises that often can be precisely specified and easily communicated in a common language. The institutions of scientific development are international and the flow of information across national borders is extensive. Neither of these conditions characterizes technology. Technological knowledge accumulates in local institutions in the form of learned know-how. It does not flow easily across national borders. In firms it accumulates in the form of skilled workers and proprietary technology, and difficult-to-copy know-how. In communities it accumulates in such diverse forms as suppliers, repair services, and information networks. It accumulates in nations as the skills and experiences of the workforce and in the institutions that train workers. Because this kind of local (or nontraded) knowledge is crucial during the initial development phase of new products, particular national solutions or advantages are not easily imitated.

Several important notions about technology trajectories follow. First, the *composition of production* in an economy matters; for long-term national development of an economy, a dollar of grapefruit production is not the same as a dollar of computer production.[34] The *future* growth and technological development of nations are molded by the *current* composition of its industries and activities. A nation's current competitive successes and failures in international trade will powerfully influence the arenas in which firms can accumulate technical skills, undertake innovation, and reap economies of scale. Since the growth and technological potential of different areas are not the same, today's outcomes set a trajectory of development.

The second key notion is *linkage*.[35] The ties between activities in an economy are not described fully by a model of market exchange, an assessment of who buys what from whom, or by a giant input–output model. Some activities are tightly linked; that is, if one disappears or expands, the second will as well. Other activities are more loosely linked. Activities are tied not just by geographic proximity but by market and community organization. For example, parts suppliers are tied to Toyota in different ways than they are to GM, with the result that product introductions are faster in GM and just-in-time delivery systems are possible. Indeed, Keiretsu supplier arrangements and financial relations create very tight technological links among firms in Japan. The way technological learning and knowledge accumulate in a national community will turn on the character of these organizational and market linkages.[36]

The third key notion, implicit in the preceding discussion, is *spillover*. Technological knowledge in one sector or activity often provides the basis for innovation in another. Knowledge spills from its sector of origin; how and where it flows is a crucial question. If the technological spillover is tacit technological knowledge flowing within a community, then that creates a distinct local advantage and sets a particular trajectory.[37]

The conclusion is that if linkages are tight and spillover effects are pronounced, the course of a few critical industries or technologies can shape an entire economy.[38] Because technologies are linked, the directions of effort and evolution are set by a cluster of "bets." The outcome—the winners among competing possibilities—emerges when the sunken investment becomes so great that radical alternatives are too pricy to adopt. Broad market acceptance of a new technology, at some point, excludes new possibilities.[39]

A nation's technological profiles—which express the composition of production, linkages among activities, and technological spillovers—emerge from a distribution of industrial and technological bets. Those bets, and the innovations they generate, emerge from a complex interaction among three factors: market demands as expressed in prices, needs that might be satisfied but are not yet expressed by buyers and sellers in the marketplace, and new additions to the "technical pool." Perhaps the best analogy is to covering the table at the roulette wheel.[40] The multiple bets that technological development requires will not be placed evenly around the table. Instead, they will cluster in two areas, according to two principles. First, research and development bets will be historically rooted. They will reflect the past development of the firm and the national economy and tend to follow the direction of past work. Put differently, the resources available for tackling the next round of technical problems will reflect what came before. Second, the needs to which technology is applied will be different in each national community, and so the tasks technology must address will vary. These two principles suggest that nations and firms within particular nations will place different bets; and those bets will become the basis of a national technology trajectory. Local variation has several sources. Differences in the industrial, institutional, or community structure can create distinct microeconomic dynamics that drive technology development along particular trajectories. The logic of each national market—and thus rational market behavior of the firms in the market—can be shown to be shaped critically by these microeconomic factors.[41] The winning technology paves a route for followers. It emerges from and plays to the national strength of the innovating country.[42]

In sum, the notion of technology trajectory emphasizes the particular and the local sources of development technology. Technology trajectory arguments refute the proposition that industrial organization is primarily the product of current technology and that national trade performance is the result principally of factor endowments. Nor does the performance of a technologically sophisticated economy derive simply from macroeconomic variables, such as savings and investment incentives. Local conditions and the history of prior technological choice matter more. Industrial organization and character are path-dependent because technological choices made at an initial point establish learning curves and create experiential know-how that cannot be bought and sold as commodities in a market. Sunk costs that go far beyond what can be measured in fixed plant and investment constrain later choices.

The New Trade Theory

The new trade theory also argues that trade is not always about mutual gain but often a competition about national futures. It argues that in oligopolistic industries governments can reshape the structure of global competition and global industry.[43] Since most major industrial sectors consist of a limited number of large powerful firms—oligopoly—the new trade theory really addresses the core of industrial competition among the advanced countries. Strategic trade models primarily examine the once-and-for-all gains to be obtained by different patterns of resource allocation, a market-determined one versus a policy-driven one. In this static orientation, strategic trade theory speaks only to the problem of allocating *existing* resources. The argument is that, under certain stringent conditions, a policy-induced outcome may generate improvements in economic welfare for one country at the expense of others. This is important, but it is not the only or necessarily the most important consequence of strategic trade.

The more consequential proposition is that promotional policies aimed at specific sectors can permanently alter the competitive balance in those industries and other industries linked to them.[44] When governments provide subsidies or protection to firms competing in oligopolistic markets, the firms make use of those increased resources to pursue different market, pricing, production, and product strategies. Not all firms will effectively use those resources to build an improved and defensible market position, but those that do will have created comparative advantage. Strategic trade policy can also backfire: policies adopted without regard to the logic of competition often undermine the very firms such policies purport to help. But there should be no question that governments can and do act to create advantage.

Strategic trade can generate extra profits for firms, but it can also benefit welfare for the nation as a whole. Oligopolistic industries tend to generate high returns (excess profits or quasi-rents). If those returns are captured by one country—for instance, if government policy wins large market shares for domestic producers in world markets—then national welfare will have been improved. The computer industry—a high-wage, high-profit industry—is a good example. National protectionist and promotional policies that capture a large share of the world computer industry for domestic producers and workers may improve national welfare at the expense of competitors abroad. Strategic trade policies act to shift the world pool of returns in a particular high-return industry from one set of national producers to another. In principle, strategic trade inherently is a "beggar-thy-neighbor" type of policy.

Of course, if resources were to move smoothly between uses without generating unemployment or economic dislocations, then one country might want to accept a foreign subsidy as a gift to its consumers. Apart from the question of why a government should allow those in the targeted industry to bear economic dislocations imposed on them by foreign government choices, resources do not move as smoothly as textbooks suggest. This is one cost to the victims of strategic trade. But the one-time costs of adjustment to imports based on foreign subsidy are not really the core issue.

As long as technologies are diffusing internationally (that is, they spill across borders and break out of particular networks and communities), each nation stands to benefit from the policies of its trading partners. A technological breakthrough sponsored by government policy in Japan benefits U.S. producers just as it benefits Japanese producers, as long as the knowledge flows across national boundaries as easily as within them. If that is not the case—if spillovers are contained within national borders—then strategic trade policies will bring the interests of nations into conflict.

The potential for national conflicts of interest in the support of high-technology industries is particularly high because knowledge does not in practice diffuse easily, and because such industries are never perfectly competitive. Investment in knowledge inevitably has a fixed-cost component: Once a firm has improved its product or technique, the unit cost of that improvement falls as more is produced. The result is dynamic economies of scale, which tend to become more pronounced with growing sophistication of technology. Markets that are based increasingly on technological competition are not made up of homogeneous products and are not subject to anything like perfect competition. Under these circumstances, government policies that promote the R&D activities of high-technology industries may win a larger share of the world returns for domestic producers and workers, and at the same time generate externalities that accumulate centrally with domestic producers and only secondarily with international producers.

In other words, both "profit rip-off" and externality rationales for policies to target high-technology industries exist. If spillovers and linkages are nationally contained, the externality rationale pushes all the more for states to pursue technological autonomy. Overall, the "new" trade theory literature provides theoretical conditions under which government policies to promote particular industries or activities because of their strategic characteristics can improve national economic welfare relative to the free-trade outcome.

To recap, since most major industrial sectors consist of a limited number of large powerful firms, the new trade theory really addresses the core of industrial competition among the advanced countries at least in principle. Strategic trade, however, is not simply a matter of the one time gains or losses that result when one government's policy assists its firms to gain share in global markets to the disadvantage of its trading partners. National position for particular firms in their markets is not the only issue, nor is the current position of one nation in the international economy its final reward. Future gains and losses in terms of each nation's dynamic potential are also at stake. A country's trade policy can affect not only its own trajectory of technology, but those of its trade partners as well.

The Implication of the Marriage

The two arguments combined yield an important implication: there may be enduring national winners and losers from trade. Competing states' development strategies then become a source of conflict in the international system. The dynamic potential of current economic activities differs. A national specialization at a given moment—efficient at that moment—may not be the basis of maximal long-term development. A nation may realize an efficient allocation of resources to specialize in those industries and activities in which the opportunities for growth and technological development are least. Another nation may be on a technology trajectory that gives it competitive advantage for long-term growth. In that case, strategic trade conflicts in new industries, or in established industries being reorganized by international competition and innovation, are not just disagreements about one-time gains. Rather, an initial advantaged position affects the long-run accumulation of skills, innovations, and knowledge that can influence the relative wealth and position of national economies over the long term, and, consequently, government strategies to shape distinct courses of national development are valid responses. In that case, trade outcomes become a source of international political conflict.

These arguments have substantial explanatory power. They help make sense of Japanese development after World War II as well as the rise of Korea in the last decade.[45] The technology trajectory arguments help to account for the diverse developments in Italy and Germany as well as in Silicon Valley.[46] They also provide a useful guide to policy choices at home and abroad.

The difficulty is that they also suggest that trading partners may adopt conflictual trade and domestic policies. Policy logic begins to look like this: if one state observes a second state taking steps (or tolerating business practices) so that foreign access to critical science and engineering is impeded, then the first state has reason to fear enduring disadvantages for product development. The first state may also be concerned that restricted access to components or subsystems made or designed abroad could create enduring disadvantages for the capacity of its national firms to conceive, develop, and introduce a given group of products. If market access for new products is also retarded, it can hinder firms' ability to establish defensible market positions in the particular recalcitrant country. These concerns exist on all sides and have long been expressed by Europeans about the United States, and now more recently by the United States about Japan and Europe as well. Theory meets practice at this point. The more theoretical arguments are accepted, the greater the perceived stakes in individual trade disputes among states.[47]

In the United States, these arguments have been debated mainly with an eye on Japan. Many Americans argue that strategic trade policy altered the course of Japanese industrial and technological development—more precisely, that trade policy was an essential feature of the particular course of Japanese development. They contend that asymmetric market access—Japanese markets being at best semipermeable, whereas American markets have been and remain mostly open—has powerfully influenced the course of technological development in the United States and Japan. By this logic, market closure supported Japanese domestic growth while distorting the development of others. Policies that promote import substitution and global advantage today can permanently change the terms of global competition and put a national industry on a distinct trajectory to enduring advantage. Even if a country's promotional policies are abandoned or changed, the global industry is permanently altered. As a consequence, policy and structural obstacles to Japan's market are seen as a real threat to foreign competitors.

To see the significance of strategic trade policies and strategies of domestic promotion, let us set aside the question of the degree to which the Japanese market has been opened in recent years and consider, rather, the role of market closure during the years of rapid postwar industrial development in Japan.[48] The thrust of developmental policy was to ensure the development and international competitiveness of Japanese producers by preventing foreign manufacturing firms from entrenching their position in the Japanese market. The effects of Japan's earlier development policy on current trade can be seen by imagining a three-phase process that, in our view, represents Japan's postwar development in a range of sectors. In the first phase, Japanese firms were at a disadvantage in both product development and production cost. Consequently, foreign firms could have dominated the markets, building up their own distribution and service systems. If this had occurred, displacing these foreign firms later would have been difficult. Tariffs or quantitative restrictions on imports would have encouraged foreign firms to open production in Japan to defend their markets. Only outright discrimination—preventing foreign firms from establishing distribution, service, and production in Japan—preserved the domestic market for domestic producers. In the second phase, Japanese firms, by borrowing technology, closed much of their product/production disadvantage. They built up distribution and service channels. Foreign firms, having lost all or most of their product/production advantage, no longer had a base for easily entering the Japanese market. Moreover, because they were largely excluded from direct contact with the Japanese market and sold products mainly through distributors, foreign firms tended not to design products for Japanese consumers or evolve production processes needed to remain competitive in the rapidly expanding Japanese market. In the third phase, Japanese producers began to build world market position. They developed distinctive products for the Japanese market that provided the basis for market entry abroad. This was certainly the case in automobiles, for example.

Once this third phase is reached, the ordinary market logic of the product cycle is at work. On the basis of distinctive products, often originally developed by Japanese firms for Japanese markets, exports begin and distribution net-

works abroad are built. Just as important, the production innovations generated by the logic of declining cost curves then gives Japanese producers real
cost advantage as well. The presence of foreign producers holding substantial
market positions would have precluded Japanese producers from driving down
those cost curves in the same way. If these dynamics hold—whether as a result
of policy or of industrial structure—it is little wonder that American companies
and government fear the long-term developmental consequences. Again, the
U.S.-Japanese trade disputes, once in traditional sectors and now in high technology, are precursors of a style of argument we are likely to see in the future
among the three major industrial regions. Certainly the Japanese developmental system has loosened and markets are more accessible. But this understanding of the Japanese past affects the way current policy and industry issues are
interpreted.

These arguments of technology trajectory and strategic trade theory create
the temptation for states to adopt a strategy of economic offensive for national
advantage, as we argue in a moment. The Japanese case is thus interpreted to
mean that market impermeability, dumping to capture foreign markets, and
intellectual property policies are strategic trade issues. We could choose other
cases from the United States or Europe. But Japan represents the systematic
national success that has forced the United States to reassess conventional theories of trade and technology. Since theory does seem to meet reality, trade policy (and more broadly innovation policy) becomes an important part of a
national development strategy that can disadvantage national competitors.
Early action to capture position in a new sector can have long-term consequences for economic development. Recapturing position can be extraordinarily difficult, as the discussion of electronics in Chapters 3 and 4 demonstrate.
Defense, it would appear, is more difficult than offense. The basis for regional
rivalry to degenerate into twenty-first-century mercantilism is present in both
the particular trade cases and the domestic justifications for action.

Regions or Nations: Does the Argument Change?

The significance of this intellectual foundation for development strategy is
amplified by the move of the world economy toward three trading blocs. Geography and tradition certainly makes us expect that trade would concentrate
into these three groups in any case.[49] Consequently, the creation of free-trade
zones in Europe and North America would be expected to divert trade only in
a limited way. Regional free-trade zones, it would seem, need not directly
undermine multilateral free trade. The critical issue is not trade diversion as
such.

However, the three regional trade blocs provide fertile ground for the
growth of strategic trade. Each contains markets large enough to influence
global competition, and each market is almost certainly large enough to capture most scale economies. Moreover, increasing trade ties within one region
will generate distinct regional industrial supply bases. That is, technology tra-

jectories will not simply rest on local or national foundations, but will reflect characteristics of the region as a whole. Linkages, spillovers, and learning will be concentrated within regions. We should expect, then, that the strategic trade game can be played with greater strength and with greater consequence from a regional base.

Let us consider, for example, how the North American free-trade talks can end up degenerating into a strategic trade game. The U.S.–Canada discussion is the simpler of the two sets of negotiations. It fits within the classical model because a United States–Canada trade zone is not going to be more closed to outsiders than was the United States or Canada alone. Trade expansion between the two countries can expand trade with their partners. The countries, as distinct as each is, have a similar level of economic development and share many legal and political traditions. Moreover, the Canadian economy (GDP) and population are an order of magnitude smaller than the United States, and the countries are each other's largest trade partners. The consequences are that for Canada, the United States is crucial in any case, whereas for the United States, Canada is simply too small as an economy or a population to create massive dislocations. Successful conclusion of these talks therefore is not surprising.

The Mexican–American (and Mexican–Canadian) talks pose a very different challenge. Mexico is the poor neighbor of a rich giant. Its economy (GDP) is roughly one-twentieth that of the United States, and its population one-third.[50] Mexico's stake in initiating these tasks is to entrench a shift from a state-oriented development strategy of import substitution to, if not a simple export-led strategy, then at least a market-oriented open trade approach. The U.S. stake is more a secondary concern of stabilizing its southern neighbor and perhaps stemming the flow of immigrants, if Mexican growth truly begins. There are real obstacles to a resolution: the Mexican economy reflects a web of subsidies that must be unwound; energy pricing to oil-using industries must be negotiated; environmental dumping and low-cost Mexican production made possible by young environmental issues are a real concern. There will be trade dislocations also, but since the Mexican economy is only the size of Los Angeles, it can't be truly disruptive to the United States.

One version of the basic bargain of the U.S.–Mexico agreement is this: Mexico will export labor-intensive goods and import technology and capital-intensive products. Thus, the American adjustment should be positive— toward high-wage, higher value-added jobs. Moreover, Mexican production could displace Asian exports to the United States. That would reduce our trade deficit with Asia, where Japanese technology holds sway limiting American exports. Mexican production could also reinforce a North American supply base in a multitude of standard components, making North America overall a more attractive production location. Mexico grows, American capital exports grow, and overall trade expands. This is the story behind the present negotiations. And it is a plausible story if we don't examine the rest of the world carefully.

An alternate, more troubling version of the tale leads us into the world of strategic trade. The first part of the alternate story is the import of labor-intensive goods to the United States, but it doesn't impact capital-intensive R&D goods from the United States. Rather, Mexico imports production equipment and advanced technology products from Asia (and secondarily from Europe) because in many sectors American producers have fallen behind. Consequently, a triangular trade investment pattern is set up.[51] The United States imports labor-intensive products from Mexico; Mexico imports capital goods from Asia and Europe, and the United States borrows to pay for its imports, recycling Asian profits. The second part of the story is that whereas Mexico's GDP is only 5 percent of that of the United States, its population is roughly 25–30 percent of the United States. Consequently, the capacity to expand labor-intensive semi-skilled production in Mexico is enormous. In this second story, Mexico grows rapidly as a production location for the assembly of consumer durables and the fabrication of simpler components and subsystems to meet North American local-content rules. Suddenly, pressure surges on a wide variety of firms using traditional production and semi-skilled labor. This scenario already has led to proposals that North America reserve the capital equipment market for domestic (i.e., North American) producers.[52] That is, the construction of the North American region could lead to development-oriented external barriers—barriers aimed at supporting American reentry in production equipment. The North American free-trade talks then suddenly become part of a strategic development policy.

The Cult of the Offensive:
Implications for International Political Economy

The simple version of this marriage of technology and strategic trade theory is that nations are competing about their economic future—that there are winners and losers. It supports a view that regional trade groups are rival blocs. The contrast is stark. The implications of traditional trade theory suggest that the mutual gains from exchange predominate; indeed, in strong versions of the theory, those who would "cheat" simply cheat themselves. In the traditional case, the international problem is to assist national governments to resist the pressures from individual groups who must pay the price for the community as a whole to capture the gains from adjustment. This is not to say that the new arguments suggest that trade relations can only be competitive. The new theories and the conventional ones are not necessarily contradictory. The traditional theory about gains from trade is at least partially valid, but it is only part of the story. When the new trade arguments are considered as well, the risks of offsetting trade strategies eliminating the possibility of capturing a privileged position or trajectory have to be set against the visible costs of open trade and internationally shared technology. Conversely, the net gains from open trade have to be judged against the risks that competitors are playing a strategic game. Precisely because development trajectories over a long period are the stakes, not just jobs or profits in the short term, the possible losses from inaction can be judged very great. Thus there is often a genuine duplicity in the actions of governments that accept both theories. The central issue is that the principles of multilateral liberalism no longer go unchallenged.

Critics of strategic trade theory argue that regardless of its possible validity, governments can rarely put these ideas successfully into practice. Economic theory may show it possible to create advantage by supporting the right industries at the right time, but politics, bureaucracies, and parochial interests get in the way of knowing what is right or acting appropriately on that knowledge.[53] A government cannot subsidize every industry, infant or otherwise, that comes looking for support with the claim of being a critical part of the supply base. And even if governments gain expertise at "picking winners," the U.S. government at least (and others like it) will continue to have a terribly hard time dropping losers.[54] Japan has probably been in a more advantageous position on this score due to its government bureaucracies' relative insulation from political pressure. That may be changing some, both because of the growing financial and technological independence of the major firms that makes it harder for government to influence their behavior and perhaps in part because of American calls for increased transparency. Strategic trade—which remains in principle valid as economic reasoning—may in practice become increasingly difficult for governments of the advanced industrial nations to implement successfully.

That apparent difficulty of implementation may not matter so much for the character of relations in the international political economy. The idea and the effort, not its successful implementation, may be what count for the international system. An analogy from security studies helps explain why.[55] Students of military doctrine, looking at the case of pre-World War I militaries among others, have identified a number of reasons why among military organizations offensive strategies tend to predominate over defensive strategies. These reasons fall into two broad categories: parochial material incentives and ideas. The power of material incentives is obvious: offensive strategies generally translate into greater influence and autonomy for the military as an institution. That took the form, for example, of larger force requirements and larger budgets. Similar incentives for influence and autonomy exist and are difficult to tame in government bureaucracies that deal with industrial strategies, even if the bureaucracies no longer function (or never did) to the net benefit of the state. *Within* each state, then, offensive strategies have a natural advantage over defensive strategies that imply less autonomy and less significance, and often lower budgets, overall for mainly self-interested bureaucracies.

Between states, the ideas of strategic trade and the behavior they activate may in the end be more important. Before World War I, a group of states each developed force postures and doctrines consistent with the idea that offense was dominant over defense—that it was easier to move forward and seize territory than it was to defend it. Wars, in this conception, would be short and decisive with a tremendous advantage accruing to the side that struck first. This set of beliefs, which Stephen Van Evera and Jack Snyder call "the cult of the offensive," turned out to be wrong. Statesmen and military leaders vastly misunderstood the weapons technologies available in 1914; the course of World War I demonstrated in practice that defense was dominant over offense. But the military and political strategies of leaders prior to the war consistently reflected the opposite belief. The immediate consequence, according to Van Evera, was a crisis that spiraled unnecessarily into war as states moved to capitalize on first-strike advantages or to prevent their adversaries from doing the same. It may be that World War I would have been fought regardless of the cult of the offensive, but the course of the war was certainly made different by the peculiar military doctrines of the belligerents.

More important for our concerns here, those military doctrines significantly affected the character of political and strategic relations between states prior to the outbreak of war. Offense seemed to export offense. States that prepared their militaries for quick decisive strikes forced their neighbors to do the same, if not to take advantage of the possibilities, then simply to deter other states from doing so. In a world where offense was believed to be dominant, even states that preferred the status quo had to protect themselves against a first-mover advantage by being ready to exert that advantage themselves. The security dilemma was at its height. Secrecy was essential, and anxious suspicion of others' behavior was rampant.[56] Preparing to preempt became prudent policy, since striking first seemed to promise inordinate advantages. Even if the advantages were uncertain, what was clear was that it would be much worse not to act while the adversary did.

The logic of strategic trade, when coupled with the notion of technology trajectories, has similar implications for economic competitors. States that make the right technology bets early on may seize critical first-mover advantages and capture whole downstream sectors of advanced economies within their borders. Others will be left behind as victims of a *fait accompli*. The language looks extreme; admittedly, there seems an odd determinism to the way we have presented this argument. In the real world, economies of scale and high entry costs are not impenetrable barriers; technological know-how is not entirely captured within national boundaries. Technology can change in unpredictable ways, leaving previous bets misplaced and surprising competitors in a favorable position.[57] There are, moreover, real gains from open trade. Yet these "realities" may not be the determining factor in how states make policy—and in any case, relations between states, economic and otherwise, do not gravitate toward some technologically determined equilibrium. States act on the basis of a set of beliefs about the world and how it works. There may be preferable solutions to conflicts of interest that are never found, because beliefs earlier have established a trajectory of relations that is very hard to get off. Certainly that was part of what happened before 1914, and it may be part of what is happening in the international political economy today.

Strategic trade can drive out competing ideas through the logic of offense dominance and first-mover advantages. Trade and industrial policy formulation in both the United States and Europe over the past few years show some of this effect. In Washington, strategic trade theory must contend with a traditional belief in free-market and free-trade approaches that remain strong and are deeply embedded in government institutions.[58] Despite this, the proponents of strategic trade—armed with a powerful intellectual and political rationale—have had an impact both on trade and on industrial policy, in the semiconductor industry and elsewhere. In Europe, the effect of these ideas has been more pronounced. It is not simply the Airbus program that is driven by this logic, but a whole range of European technology programs. Indeed, this concern is entangled directly with the reemergence of a drive for a unified European market. (See Chapter 3.) These European undertakings coexist, of course, with real efforts to reduce trade barriers and industry subsidies and dismantle "inappropriate" business–government liaisons. Indeed, the most recent technology development efforts in the Community aimed at reestablishing the European position in advanced electronics are now expressed in the liberal language of market competition. In Japan, rhetoric—and to some extent practice—has evolved as the developmental model has loosened. However, commitments to ensuring Japanese competitive position in advanced industry have not weakened.

What kind of international system might emerge if these notions are widely believed and embedded as assumptions or principles in the economic system? Can a variant of managed multilateralism based on modified free-trade principles as the model for future trade and financial relations survive the theory and practice of strategic trade policy? The future balance of policy and principle is by no means clear in any of the three regions and will be the result of both domestic and international policy conflict. Alarmist views sometimes extrapolate to the possibility of overt conflict, even military conflict, between the great centers of advanced industrial capability. That goes much too far. The use of force among these countries is restrained by norms, by the intense destructive capability of modern armies, and ultimately by the existential deterrent that nuclear weapons provide.[59] Force is also devalued by the nature of modern economies: The kinds of economic activity that are now prevalent and are likely to be the leading sectors in the future are less tied to physical resources and are less easily appropriable within geographic boundaries. There is little reason to think that economic competition among advanced industrial countries could in the foreseeable future overwhelm the peaceful resolutions that have characterized their conflicts of interest since 1945. That said, security could nonetheless reemerge as a fundamental question among states that for nearly half a century have been able to cooperate and compete without the impediments and dangers of viewing each other as potential adversaries.

It is also naive to reason in the opposite direction from nonuse of force among states to a liberal world economy or even a heavily compromised, managed multilateralism. From what we know about the connections between economics and security, we deduce that an environment in which the use of force is restrained is a prerequisite for the higher levels of interdependence that come with liberal economic relations among states. But as a prerequisite it is only a necessary and not a sufficient condition.[60] A neo-mercantilist world in which states or regions seek economic autarky and compete for advantage vis-à-vis others in domestic welfare *and* state power is equally plausible. That could incite new economic and political tensions, which would dramatically change the character of relations among the advanced industrial states, as well as between each and its semi-periphery and periphery. There would also be important implications for life in the parts of the world that are left out of the system.

How could the three trading groups come to view each other as potential adversaries? There are two distinct mechanisms at work that could produce this outcome. One conventional view is that disagreements over trade issues will lead to greater protectionism among the United States, Japan, and Europe; and that the resulting tussles would engender political tensions that reinforce economic reasons for each center of industrial power to lean toward its own backyard and emphasize its parochial interests there. That would certainly be to the detriment of any continuing multilateral cooperation between them, and the effect would extend beyond just trade. In issues as diverse as global environmental change, contributions to aid in the least developed countries that are not a vibrant part of the world economy, and so forth, cooperation would become increasingly difficult. Consensus over broad conceptions of security would also suffer, particularly given the lack of any single convincing threat to replace the Soviet Union. Tensions could arise over new security issues that would energize the interests of each regional power quite differently.

The "best" outcome in such a world of defensive protectionism would be a version of benign regionalism, whose characteristic feature would be disengagement and minimal relations among the three blocs to avoid the strains of direct competition. In areas where disengagement was prohibitively costly, benign regionalism would have the three centers of power handling their differences through an agreement, more or less, to disagree. Such an outcome would still rest on the ability and willingness of the three blocs to substantially reduce the level of interdependence among themselves. Each bloc would work to develop its own supply base and cultivate its own markets, and the world economy that we know today would divide essentially into three subeconomies. It is possible that three large and inwardly oriented economic blocs could coexist comfortably—subject to a shared understanding that the drive for autarky came from mutually recognized desires for regional autonomy, control over domestic welfare, and internal political stability, not competing state or bloc power. But it isn't likely. The extent of connections in the international economy is too great for the regions to fully separate from each other, even if their degrees of freedom are much greater than is usually understood. The more likely result is that those interconnections will produce fears of exploitative dependence and strategic moves to avoid, or more insidiously to create, such dependence.

Strategic trade and technology trajectory arguments provide a powerful intellectual and political rationale that will tend to bolster worries about relative position among the three blocs and reinforce the notion that economics is still about competing state power, not just generating riches for consumption.

The tilt toward neo-mercantilism could become more pronounced as the areas in which the blocs compete for strategic trade advantages move closer to the heart of military technologies. In the abstract, each bloc could in due time probably develop the autonomous capability and expertise necessary to field and supply a modern military force, without dependence on others. There are already pressures in Europe and Japan to move toward independence. At the same time, the United States is moving toward greater dependence. The possibility of American dependence on access to standard weapons systems components that are made offshore has been a concern and speculation in Washington for some time. At an extreme, there is the concern that weapons or components could be withheld when the security interests of a trade partner differ from those of the United States, and there exist fears that this may already have happened in small but visible ways.[61] Particular incidents that suggest dependence and its consequences, will certainly bolster arguments for autarky and will prod the United States to try to bring as many relevant capabilities back onshore and design its future weapons systems to make use of what is available at home. At a minimum, this would lead to further disengagement and fervent protectionism. Unfortunately, the nature of links between commercial and military branches could make the outcomes worse.

Consider the implications of current Japanese dependence on American and European space-lift capabilities for launching communications satellites essential to Japan's consumer television networks.[62] That system is now in considerable trouble, following the explosion of two launching vehicles during attempts to put new satellites in orbit in 1990 and 1991. The first of those launch vehicles belonged to Arianespace, and the second to General Dynamics. Debacles like these give support to voices in Japan that argue for greater technological self-sufficiency. In the specific area of rocketry, Japan's own heavy-lift rocket (the H-II) is not expected to be ready before 1993 at the earliest, but it and follow-on programs are almost certain to receive a boost from the recent events. Regarding satellite technology itself, in 1990 Tokyo made a substantial concession to the United States in agreeing to drop government-sponsored development of advanced satellites and open bidding to foreign competitors. In principle, that would seem to be a move toward increased interdependence, but it may not work out that way in practice. There recently have been problems in the power supply of the primary satellite that beams programs of the semi-official broadcast agency NHK to at least 4 million households in Japan—and that power supply was made by General Electric. Although this problem is unrelated to the rocket failures, they combine in a highly visible way to demonstrate to a wide audience the immediate costs of technological interdependence. That is particularly significant in areas such as rocketry and satellite systems since the technologies and industries that are critical to these sectors are also prime candidates for strategic manipulation by governments. They are also, importantly, closely related to military technologies and military systems expertise.

The changing nature of links between commercial and military technologies makes the problem of ensuring technological autonomy all the more complicated. Government can not simply cede commercial markets and still develop critical military technology products. Spin-on applications of commercial technology to military systems means that states will now more frequently than in the past be choosing subsets of civilian commercially developed technologies to develop into military applications. In other words, it is now the civilian sector that frequently will generate the supply base of skills, product knowledge, process know-how, components, and subsystems required to produce the military products. Without the related commercial industry, it will not be possible to produce the specialized military goods. But not all commercial technology will, in fact, be spin-ons. Because the range of commercial technologies in critical sectors such as materials, aerospace, and information processing is so wide, new weapons systems will incorporate only a small fraction of what is possible. The designers of the new weapons systems, or the officers who guide R&D for the military, will have to make choices about which of those possibilities are in fact the most promising. They will choose which civilian technologies to work with and which to leave aside. Those choices will "place the bets" for military systems and will affect the trajectory of future possibilities, just as happens in the commercial sector. Precisely because choices made for the military are linked naturally to bets about what technologies are most promising in the commercial sector, there will be abiding pressures to keep the choices secret. States—even putative allies—that support research in precompetitive generic technologies in military sectors will have a clear incentive not to share even with their allies information about what technologies they have chosen to push.

That is an insidious consequence of the logic of the economic offensive. Absent that logic, it made good sense for the United States to collaborate with the Japanese and the Europeans in the Strategic Defense Initiative research program, including some of its more high-technology and leading-edge aspects. But if states act according to the arguments of strategic trade or believe that others are likely to do so, this kind of collaboration will be heavily disfavored—simply because revealing military technology bets at early stages of research could have important consequences for commercial technology bets. Because the decision to place bets on one or another technology is the most critical part of creating advantage, states that practice strategic trade have the strongest incentives to keep those bets secret. A state that believes others practice strategic trade has comparable incentives, even if the state itself has no intention or would prefer not to pursue strategic policies. At the same time, states will worry about which technologies other states have chosen to work with and why.

Great secrecy in military research and development at this level may not be compatible with alliance and continuing security cooperation over the long term. The controversy that developed over the FSX fighter plane, bitter as it was, did not extend to uncertainties about what the Americans and the Japanese had at stake, and what technologies they were in fact trying to protect. If that were different in the future, similar controversies might have much more serious consequences. Estimates of political intentions aside, cooperation between states is difficult to sustain when the cooperators cannot look out into at least the medium-term future and agree broadly on how their interests are likely to be affected by what they do together. If states do cooperate under such circumstances, failed expectations are apt to generate greater tensions between them and complicate future relations.[63] All of this would work against the quasi-philosophical foundations of a benign regionalism, undermining perceptions that economic competition could to a substantial degree be divorced from state power.

We do not believe that a new mercantilism would recapitulate the past. But we do foresee possibilities for a great deal of tension to develop between the blocs, and that tension would not necessarily be confined to economics. This will make it extremely hard to deal with global issues that could otherwise be subject to resolution via consensus among a few powerful states, as the optimistic view of managed multilateralism foresees. Many things that could have been done by cooperation among a committee of three, that did get done earlier while the United States was basically a hegemon, will not get done in this less happy vision of a multipolar world. Major war, seemingly almost impossible in the foreseeable future, may very well be ruled out even in the long term. But the character of international life for both small and large states in this neo-mercantilist world could be significantly different from what we have come to expect during the past forty-five years, and what we might aspire to now that the Cold War is over and the technological potential for great wealth is expanding in fabulous measure.

Conflict or Collaboration?

The economic foundations of the old security system eroded with the emergence of Europe and Japan as industrial giants and the accompanying redistribution of national capabilities. The character of the international economic system that is now being constructed will, in the absence of common direct and clearly defined security threats, powerfully shape the overall system of international relations that will emerge within that new structure of capabilities. The world economy now consists of three largely self-contained trading groups, despite supposedly "global" companies and international financial markets. The open question is what the relations among those three groups will be. We cannot yet answer whether each trade group will become an inward-looking bloc in a system of tense rivalry emphasizing particular gains or become a supportive partner in an open interconnected system of mutual advantage.

The three models—managed multilateralism, benign regionalism, and twenty-first-century mercantilism—are not really alternatives. Aspects of each will be evident in the reality that unfolds. Managed multilateralism suggests a network of global finance with careful central bank collaboration to maintain stability and the continuing expansion of direct foreign investment that links operations of global companies across the three regions. Benign or protective regionalism is evident in the automobile quota systems set up in both the United States and Europe to contain Japanese imports in order to limit social dislocation. Twenty-first-century mercantilism is evident in the trade debates about semiconductors and aircraft in which the position of national industries in advanced technologies is the stake, and the role of government in promoting that advance is at issue. The character of the system will not be defined simply by counting and weighting aspects of the economy or disputes that fall into one category or another, though the balance will be important. The character will also be defined by the principles built into and enforced by international agreements and institutions and by the conceptions of the actors about the constraints and possibilities of the system, though of course not simply their words. Those principles and conceptions that define the system, moreover, will emerge from critical conflicts as much as they will create the outcomes of those conflicts.

An economic order of managed multilateralism is almost certainly the best foundation for a cooperative security system. It would ensure that economic competition would not become the basis of or contribute to rival conceptions of security on the one hand, or to a distrust about intent that is then expressed as concerns about security on the other. Japanese and European technological prowess and wealth need not be translated into a full-blown military autonomy; and if it were, that autonomous capacity need not be deployed for purposes opposed to those of the United States. Managed multilateralism becomes a cushion within which security relations are organized. That cushion is necessary now that threats are less immediate and more varied; this is in real contrast to the focused and immediate threat of the postwar era when the security structure cushioned and supported the economic system. But managed multilateralism will not emerge automatically, if it emerges at all, and an American position of leadership in such a system is hardly assured.

To establish managed multilateralism as a foundation for new security order, unconventional issues must be resolved. Traditional international economic institutions and conceptions in all likelihood will not be able to resolve the extraordinary issues. Consequently, managed multilateralism will require the creative evolution of established organizations such as the GATT and likely the creation of new institutions to buttress managed multilateralism in a new era. A failure to resolve them would certainly risk an era of mercantile rivalry.

Epilogue

MICHAEL BORRUS, WAYNE SANDHOLTZ,
STEVE WEBER AND JOHN ZYSMAN

The New Security Game

The end of the Cold War not only gives us an opportunity to redefine the international security system, but also makes it necessary. As this book goes to press Soviet communism lays in ruins, the Baltic states are independent, Eastern Europe is once again part of Europe, and Yeltsin's Russia is struggling to reorganize the Union and build a democratic market economy. The United States appears to have achieved many of the goals of the security system it constructed almost fifty years ago. Yet, even as the threat of tanks rolling into Western Europe recedes, the problem of intraregional migration intensifies.[1] Ethnic instability waxes even as great power confrontation wanes.[2] The "end of history" is nowhere in sight; the character of the security system must change to reflect new challenges.[3]

As the impetus behind the post-war security system fades, the concerns emphasized throughout this book gain even greater currency. A new economic foundation is emerging for the next security system. In the old system, with capabilities outside the Soviet bloc concentrated in American hands, there were relatively few constraints on U.S. policy. At this historical moment of American triumph over communism, U.S. political credibility is still high. But the new order will be formed at a time when there are sharp constraints on U.S. policy and with a broad distribution of resources across the three major regional groups. Europe and Japan have the economic power to challenge American leadership. Now that a common and dominant Soviet enemy no longer exists, their priorities are likely to diverge from American preferences.

This is already being amply demonstrated by European policy and concerns in Central Europe. Although the United States counseled restraint and caution, Europe rushed to recognized Baltic independence. Europeans have also taken the lead in providing financial support for Central European political stability and in trying to resolve the divisive conflicts in Yugoslavia. They are even flirting with a European intervention force. Diverse security interests pursued independently in the three regions will be a fact of life for the next several decades.

Indeed, relative to its allies, the United States is confronted with a diminished capacity to generate and deploy economic resources in pursuit of foreign policy objectives. Where there was once leverage in finance, trade, and technology, there is now constraint. U.S. capabilities to shape discussions and outcomes—through cajoling or coercion—are now fundamentally circumscribed by allied willingness to comply with U.S. goals. American financial influence is altogether different than, for example, in 1956 during the Suez crisis, when the United States was able to halt a British, French, and Israeli invasion in part by threatening a run on sterling. Today, Germany and Japan have the economic capacity to constrain U.S. policy should they so choose. America's hat-in-hand appeal for Gulf War financing has made it clear that U.S. influence is largely dependent on allied toleration of U.S. aims.

The United States is similarly constrained by emerging industrial weakness and a declining ability to ensure access to its domestic market in return for compliance with U.S. foreign policy. In traded industries today, U.S. policy protects an increasing variety of competitively troubled American sectors, and other countries look to emulate the Japanese model of production rather than U.S. practices. Where once the United States was the only significant source of advanced technologies, the domestic economy is now increasingly dependent upon foreign supplies of many of the component, machinery, and materials technologies that underlie civilian and military production. At best, as in the Toshiba Milling Machine case, emerging constraints (e.g., dependence on imports of key intermediate inputs such as memory chips made by Toshiba) make it very difficult to punish clear contraventions of U.S. security interests. At worst, they leave U.S. authority considerably diminished and U.S. freedom of action susceptible to the leverage of foreign power.

Complementing these constraints on the United States are the new techno-industrial and policy capabilities created by our allies and examined throughout this book. Fundamental challenges to U.S. capabilities are posed by Japanese innovations in production and technology for high-volume markets, and by formidable European industrial strengths in industries from chemicals to aerospace. Their new economic capabilities presage new international political influence. The United States continues to lead for the moment, but only by default.

The new distribution of economic strengths and resources will not determine the precise shape or character of the new international system. The realm of the possible is being redefined, but the outcomes remain contingent on the beliefs and actions of the major protaganists. The constraints and opportunities facing them have changed, but their responses matter most.

In our view, the character of the security system will turn on the complexion of economic relations among the advanced countries. Market ties have long been thought to limit conflicts, but that may be less an inherent result of interdependence than an artifact of successive free-trade hegemonies. The same argument was heard, after all, before World War I. Economic interconnections might continue to ensure enduring peace among the western industrial countries, but the frictions of economic and policy disputes could also accumulate into more central conflicts.

That threat of conflict increases as the world economy evolves into American, Asian, and European regions, together accounting for above 70 percent of

global GDP, and separately driven more by internal than interregional developments. The collapse of the Soviet Union does not change the significance of this
emerging economic regionalization. The old Soviet Union would have been the
least economically significant of the regions. Similarly, if a new Russia-centered
association of republics develops into a fourth region, it will be much poorer
than the others, smaller, and far less dynamic than Japan with its Asian cohort.
It would not be a direct shaper of events in the economic arena, but its choices
would matter indirectly as its alliances shifted. Alternatively, the new association of republics might integrate into the West, dramatically increasing the
importance of Europe and making Europe even more of a player and creating
internal pressures that inevitably would alter the shape of Europe's external
ties. Finally, even if the new association of republics remained largely peripheral, it would still influence events at the margin, tilting the resolution of economic issues among the three major regions. In any of these scenarios, the core
set of concerns raised in this book would remain at the center of any new international system.

Regional industrial opportunities and conflicts will shape three possible ide-
als of an international system: managed multilateralism, regional coexistence,
and regional rivalry. Regional mercantilism, the least desirable outcome, could
occur by default since neither international bargains nor institutions are necessary to animate it. The vocabulary and ideology of twenty-first-century mercantilism are already emerging under the guise of theories of technological
development and strategic trade. Under these circumstances, the persisting tensions and conflicts of an open trading system could degenerate into an outright
struggle for position and advantage. The implication for America is clear: managed multilateralism will not be maintained by accident or default, only by self-
conscious design.

Inadequacy of Old Approaches

Mercantilistic trade wars are not inevitable, but traditional approaches to containing them are probably insufficient. The institutions, ideas, and procedures
that supported the expansion of open, multilateral trade in the last half-century
are inadequate for two major reasons. First, it is less and less likely that the
United States will be able to stabilize an open world economy by resisting
domestic pressures for closure against foreign competitors. Second, the industrial practices and policies associated with aggressive technology development
and strategic trade cannot be effectively addressed by existing multilateral institutions such as the GATT. Each issue is worth a closer look.

Increased defensive protectionism in the United States could trigger a trade
war as easily as structural Japanese barriers or a Fortress Europe. Free-trade
systems create inevitable domestic economic adjustments with concomitant
political pressures to resist them. When the U.S. economy was much stronger
than others, domestic political pressures could be bought off and the commit-

ment to an open system maintained. As the competitive position of U.S. industries weakened, however, the adjustment problems of displaced American producers and disrupted communities grew. In order to maintain apparent
compliance with formal GATT rules that generally prohibit explicit trade
restrictions, U.S. policy responded to domestic political pressures by negotiating bilateral "voluntary" limitations on the quantity of permitted imports in
the trade-impacted industries.

Such informal quotas create perverse incentives that exacerbate the very
problems they are intended to solve. Since limits on imports are established,
foreign producers can maintain profits only by innovating and moving up-mar-
ket to sell higher-valued goods. Domestic price levels rise. Because profitability
has been restored at the newly higher domestic price levels, the troubled
domestic producer is no longer under pressure to adjust competitively. Meanwhile, production tends to spread to other countries since the quotas can be
avoided by exports from locations not covered by the bilateral agreements. In
short, U.S. policy has tended to encourage foreign producers to become even
more competitive, defer the adjustment of American producers, and invite new
entrants who ultimately intensify the adjustment problems.

More recently, as the dollar weakened, foreign competitors responded to
the threat or actuality of trade barriers by assembling product in the American
market. Trade disputes are giving way to arguments about foreign direct
investment (FDI). As with past American FDI in Europe, the central policy concern must be whether superior production and product know-how will diffuse
widely from the investor into the host economy. The hope of the host is that
new processes and ideas can be learned; the fear is that FDI will create an
advanced foreign enclave with limited links to the local economy. When American multinationals invested in Europe, technology and production paradigms
diffused to the benefit of the local economies. However, we cannot assume that
similar benefits will accrue from Japanese investment given the different structures and practices of Japanese industry. What political limits on Japanese success will be established in the United States? How strident will be the domestic
politics behind those limits?

The unintended consequences of past U.S. trade policies, the pending disputes over FDI, and the continued competitive problems in advanced industrial
sectors all carry the same implication: U.S. policymakers will find it increasingly difficult to maintain an unwavering commitment to a fully open world
economy. At best, increasing closure at home will substantially diminish U.S.
credibility in fighting for openness abroad. At worst, closure at home will invite
reciprocal retaliation abroad and touch off mercantilist rivalry.

The prospects for an easy transition to managed multilateralism are diminished to an even greater extent because existing multilateral institutions such as
GATT are ill equiped to deal with the practices of strategic trade and technology development. In principle, each regional economy wants access for its
firms to markets, technology, and investment opportunities in partner
economies. But Europe and Japan—and eventually the United States if our

argument here is right—also want considerable control of foreign behavior in domestic markets. That control may come by government restriction (as in Europe) or by sharp limits on the ability of outsiders to buy into existing groups or participate in market privileges (as in Japan). Mutual desires for unrestricted access abroad while maintaining restrictions on foreign practices at home are obviously hard to reconcile.

The contradictions are at the heart of contemporary disputes over trade and investment, and the disputes are not about the formal principle of freedom of access. Rather, they are about the impact of domestic policies and business practices that are not transparent and that implicate desires for national and regional autonomy in development. For example, domestic policies to accelerate productivity by supporting technological innovation are difficult in practice—perhaps impossible—to distinguish from strategies to create market advantage. Policies to promote the domestic diffusion of technology easily emulate mechanisms that support domestic producers at the expense of imports. In technology industries characterized by scale and learning economies (in which costs drop dramatically as volume increases) forward-pricing strategies are in practice indistinguishable from dumping.[4] Policies and practices that limit foreign access to domestic markets become means of preventing foreign competitors from capturing gains in an expanding national market and reserving those gains for domestic producers.

Judging whether such domestic practices and policies are simply idiosyncratic or are illegal barriers to trade requires inquiry into intent and effect. Competing national approaches ultimately force trade partners to negotiate and balance their interests. Negotiated "balancing" of interests is, of course, a euphemism for managed trade.[5] Whether considered a horrifying violation of the logical and ethical premises of free-market capitalism or a commonplace reality with its own logic and problems, managed trade is essentially about which domestic rules and market activities will be the reference in international practice. Since domestic practices are seen as inherently legitimate by each nation, negotiations end up allocating markets when domestic practices cannot be reconciled. The result is informal shifting policies and agreements that provide at best temporary relief and are always subject to change through later negotiation.

For example, United States–Japan semiconductor trade agreements are explicitly about market share and about what ought to count as acceptable global business conduct. Practices such as market allocation and dumping that would be blatantly illegal in the United States helped to propel the Japanese industry to world leadership. The United States is therefore seeking agreements that would restrain some Japanese practices while creating a target domestic Japanese market share for U.S. firms based on some judgment of the share American firms would have captured absent impediment and predation. Needless to say, these are difficult judgments to make, and they change as competitive strengths shift. Similarly, the ongoing dispute in aircraft focuses on the appropriate roles for domestic (or regional) government policy in the world economy. The American commercial aircraft industry benefited early on from defense-related development of jet engines and airframes. The dominance of U.S. producers on commercial markets was therefore viewed by European allies as the illegitimate creation of an implicit industrial policy. European governments intervened in their own economies to set things right. In turn, Europe's direct commercial support of Concorde and then of Airbus struck U.S. policymakers as inappropriate direct intervention in civilian markets.

As governments become directly involved in allocating market share and determining what industrial activities will occur in which region, trade outcomes increasingly appear as the result of political bargains rather than as the legitimate result of competitive markets. It is simple neither to identify nor even to define a legitimate outcome. The problem is not simply that it is difficult to negotiate multilateral rules specifying appropriate conduct—though that will be hard. And the problem is not only that most of the practices that need to be regulated are simply not transparent. Rather, as disputes accelerate with the practices of strategic trade and technology, a plethora of drawn-out, ambiguous, and bilateral resolutions make it impossible to give general principles consensual meaning. Formal GATT rules prove utterly insufficient to resolve the substantive ambiguities in a timely fashion. In the absense of agreed-upon multilateral principles and the institutions to enforce them, bilateral strategic manuevering will become the principal tactic, and regional rivalry the paradigm of the international system.

Toward Multilateralism

Managed multilateralism will not emerge automatically, if it emerges at all, and an American position of leadership in such a system is hardly ensured. Instead, to avoid rivalry, the three regions need to: (1) agree on a set of principles that endorse reciprocal access to regional markets, investment opportunities, and supply base technologies; (2) delimit appropriate industrial practices and policies; and (3) develop new multilateral institutions for coordinating bilateral regional moves. Each is worth a brief look.

Reciprocal access to markets, investment opportunities, and underlying technologies are the only possible paths for multilateralism wherever the temptation to strategic trade practices exists. When know-how and markets for new technology cluster regionally, and progress is driven by scale and learning, whoever has the broadest access to all three regions will likely end up dominant. Or, to put it another way, if country A has access to three-thirds of the world's storehouse of technologies relevant to our industry and country B has access to only two-thirds, over time country A is likely to win. Reciprocity of access permits as much openness as each regional economy can tolerate politically, and forces compromises in domestic practices that impede access whenever domestic industries seek foreign market opportunities.

The principle can be enforced, however, only by reaching some degree of agreement over which domestic business and policy practices are appropriate international conduct. Multinational codes of behavior are very difficult to negotiate and historically problematic to execute and enforce. Rather than itemizing individual industrial behavior, it will probably be necessary to limit the impacts of disruptive strategic practices by directly negotiating the interests involved. That is, wherever strategic practices are alleged and proved, resulting market shares of the advantaged industry ought to be limited by agreed rule of thumb. For example, at least half of domestic consumption, but no more than half of global production, ought to be produced locally (with full local value added). Foreign direct investment, monitored to ensure compliance, would be the vehicle to adjust market shares. This would bring significant local production back into an economy that had been disadvantaged by the strategic practices of its trading partner and would still reward innovating industries, but it would simultaneously help to achieve real interdependence among the regional economies.

The complex arrangements necessary to achieve these goals will be very difficult to formulate at a multilateral level. As the Structural Impediments Initiative talks with Japan have demonstrated, it is hard enough to bargain over the practices and impacts of *two* distinct economic systems, let alone three or more. This means that trade talks will have to proceed on a bilateral basis among Europe, Japan, and America. The task will not be to end these bilateral talks, but to put them into a multilateral context with rules of procedure and sufficient transparency to ensure that those who are not direct participants can still make their needs and interests felt. GATT could act as such a forum if ongoing bilateral negotiations could be monitored and a limited number of selective interventions could be exercised by individual countries or regions. GATT could also be the forum to adapt the Generalized System of Preferences (GSP) to the new reciprocal paradigm to help promote economic development abroad.

External agreements will prove ineffective, however, without reestablished strength at home. The American position in a new international order, and indeed within our own region, depends on whether an expanding and innovative American economy is a leader in production and product technology. That can only be accomplished at home. Other analyses have specified the necessary changes in domestic policies and business practices.[6] This is not the place to repeat their prescriptions in detail. Suffice it to say that any domestic economic agenda must: (1) ensure cutting-edge development of technology and best industrial practices at home; (2) develop the mechanisms for the diffusion of advanced technology and practices throughout domestic industry; and (3) provide a capacity to absorb technology and best practices developed abroad. The standards are not abstract; they are what is required to be successful in global competition.

Aggressive domestic development policy and practices need not interfere with the broader multilateral goals; indeed, they may be necessary merely to ensure that the broader goals are achieved. Here, the analogy of security (specifically, arms control policy) is appropriate. We would never have negotiated for arms control the way we currently negotiate for trade concessions—by adhering only to abstract principle and disarming in pursuit of it. Instead, we engaged in a massive arms build-up and then reciprocally negotiated concessions. Some version of this is likely to be the only effective approach in the current trade climate. Only by engaging in some systematic practices to create domestic advantage will we be sufficiently armed to reciprocally bargain away strategic practices and policies to ensure global equal access. Indeed, it was precisely the mutual concession of reciprocally lowering tariff barriers that drove the success of the first six rounds of tariff reduction in the Multilateral Trade Negotiations (MTN) under GATT auspices through the 1960s. As strategic trade practices accelerated abroad in the 1970s and 1980s, the MTN floundered in part because the United States had few such practices to concede in return for the concessions it wanted from others.

The stakes may be as high now as when the GATT was conceived. Now, as then, the stakes are not confined simply to the future of industrial competition and the comparative levels of wealth in Japan, the United States, and Europe. Then the postwar distribution of capabilities ensured American preeminence but left open the terms on which the United States would exercise its power. Now the leadership possibilities and the terms of power both are far less bounded. There could be broad cooperation among the three centers of industrial power or dangerous competition that degenerates into security confrontation.

The redistribution of potential military power, if not yet its embodiment in missiles, planes, tanks, and ships, is not the only source of possible conflict. The nature of threats has changed. Other economies increasingly are in position to exploit terms of trade to impose dominance, to structure and play the international system through economic means. The end of the Cold War threw the international system open to redefinition. But the redistribution of economic power was already shifting the foundations of the old bipolar system.

Japan and Europe are not on the verge of striking out on their own in security affairs, but they increasingly are in position to be far less dependent on the United States for defense—if they so choose—and perhaps, through economic means, even to impose limits on U.S. ends. The Cold War alliance was based on three elements: a Soviet threat, American hegemony, and allied dependence. Each element has changed. In the new world order, Japan and Europe can go their own ways not just because the Soviet threat is gone, but also because they have the techno-industrial capabilities for doing so. U.S. leadership is no longer a matter of necessity, but of political choice.

We are not arguing that Japan and Europe are about to cut their defense ties with the United States. Rather, the costs of altering their security or foreign policy strategies have been tremendously reduced, whereas the costs of maintaining American autonomy have significantly increased. In the Gulf War, Japan and Europe followed Washington's lead—with varying degrees of boldness or alacrity. Nothing would necessarily persuade or induce them to follow in a future conflict if their aims differ from those of the United States.

Indeed, quite the reverse is possible: For reasons of financial need or technological dependence, the United States might be induced to follow their lead when that was not in American interests. As we said at the outset, America needs to act not from the belief that we are and can remain dominant, but from an understanding of how we can be effective in circumstances in which we no longer are.

Notes

Chapter 1

1. This is the position essentially argued by Paul Kennedy, in *The Rise and Fall of the Great Powers: Economic Change and Military Conflict from 1500 to 2000* (New York: Random House, 1987); and, analytically before him by Robert Gilpin, *War and Change in World Politics* (New York: Cambridge University Press, 1981). Many of the forces they depict are at work in the American case, but those forces are not central to the deterioration of the American position.

2. This problem is very well analyzed by Theodore H. Moran, "The Globalization of America's Defense Industries: Managing the Threat of Foreign Dependence," *International Security*, Vol. 15, no. 1, Summer 1990. Again, much of Moran's analysis is directly apposite; it is his presumption of globalization and interdependence with which we take issue. For a fuller explication of our difference with Moran, see pages 28–29 of this chapter.

3. Samuel P. Huntington, "Coping with the Lippman Gap," *Foreign Affairs, America and the World 1987/88*, Vol. 66, no. 3.

4. Japanese Ministry of Finance data, as reported in *JEI Report*, No. 12B, March 23, 1990, p. 5.

5. Of course, this does not mean that the American economy has become like that of Mexico or Brazil. Our debt is enormous, but it is still a small fraction of our GNP, and because that debt is almost exclusively denominated in dollars, we fundamentally control its disposition.

6. That the various interpretations do not disagree over the fact of decline is amply demonstrated by David E. Spiro and James A. Caporaso, "Why Honest Theorists Disagree over Hegemonic Decline." (Paper prepared for the American Political Science Association annual meeting, San Francisco, CA, Aug. 30–Sept. 2, 1990.)

7. *Ibid.*

8. See Gilpin, *War and Change*. Also, Hans Morgenthau, *Politics Among Nations* (New York: Knopf, 1951).

9. Although nuclear deterrence has eroded the efficacy of direct territorial control, it has not upset the security goal of preserving an internal community from external influence.

10. Here we rely on the definitions of power elaborated by James A. Caporaso and Stephen Haggard, "Power in International Political Economy," in Michael Ward and Richard Stoll, eds., *Power and International Relations* (Boulder, Colo.: Lynne Reinner, 1989); and Spiro and Caporaso, "Why Honest Theorists Disagree."

11. See, e.g., Stephen Krasner, "United States Commercial and Monetary Policy: Unravelling the Paradox of External Strength and Internal Weakness," in Peter J. Katzenstein, ed., *Between Power and Plenty: Foreign Economic Policies of Advanced Industrial States* (Madison: University of Wisconsin Press, 1978).

12. See, e.g., Charles Maier, "The Politics of Productivity: Foundations of American International Economic Policy after World War II," in Katzenstein, ed., *Between Power and Plenty*.

13. This occurred normatively as well, as economic success in Europe and Asia led to the widespread adoption of U.S. culture and acceptance of many of the basic principles underlying open trade and finance.

14. See, e.g., Moran, "The Globalization of America's Defense Industries," and Kennedy, *The Rise and Fall of the Great Powers*; Joseph Nye, *Bound to Lead* (New York: Basic Books, 1990); Robert Keohane, *After Hegemony: Cooperation and Discord in World Political Economy* (Princeton, N.J.: Princeton University Press, 1984); and Henry R. Nau, *The Myth of America's Decline: Leading the World Economy into the 1990s* (New York: Oxford University Press, 1990).

15. There is a real and fundamental intellectual risk in posing problems as competing ideal types. The essence, then, is that one or the other phenomenon dominates reality, obscuring the continued importance and nuances of the other. For example, modernization theory often hides the continued importance of traditional culture by asking whether a society is traditional or modern—thereby anticipating that the traditional elements will disappear as the modern emerge. Barrington Moore re-posed the question by asking how a particular form of modern society is shaped in a particular traditional setting. His question creates intellectual freedom from the old dichotomy whether or not his particular answer is accepted. See Barrington Moore, *Social Origins of Dictatorship and Democracy* (Boston: Beacon Press, 1966).

16. This section draws on much of the work done at the Berkeley Roundtable on the International Economy (BRIE), especially collaborations with Stephen S. Cohen, and the work of Laura Tyson. See in particular Stephen Cohen and John Zysman, *Manufacturing Matters: The Myth of the Post-Industrial Economy* (New York: Basic Books, 1987); "Manufacturing Innovation and American Industrial Competitiveness," *Science*, Vol. 239, March 4, 1988; and "Business, Economics, and the Oval Office: Advice to the New President and other CEOs" [co-authored with Samuelson, Kearns, Young, et al.], *Harvard Business Review*, Vol. 66, no. 6, November–December 1988. Relevant works of Laura Tyson include "Business, Economics, and the Oval Office: Advice to the New President and other CEOs" [co-authored with Samuelson, Kearns, Young, et al.], *Harvard Business Review*, Vol. 66, no. 6, November–December, 1988; "The U.S. and the World Economy in Transition," BRIE Working Paper No. 22, 1986; "Creating Advantage: Strategic Policy for National Competitiveness," BRIE Working Paper No. 23, 1987.

17. U.S. Congress, Office of Technology Assessment, *Paying the Bill: Manufacturing and America's Trade Deficit*, OTA-ITE-390 (Washington, D.C.: U.S. Government Publications Office, 1988). This is the position argued in Cohen and Zysman, *Manufacturing Matters*, and foreseen in Laura Tyson and John Zysman, eds., *American Industry in International Competition* (Ithaca, N.Y.: Cornell University Press, 1983).

18. Overall, as an MIT study affirmed, the troubled sectors together represent a broad range of industries, often the entire industrial complex within those industries, and a substantial chunk of total U.S. output. See Michael L. Dertouzos, Richard K. Lester, and Robert M. Solow, *Made in America: Regaining the Productive Edge* (Cambridge, Mass.: MIT Press, 1989).

19. On the group structure of the Japanese economy see, e.g., Michael Gerlach, "Keiretsu Organization in the Japanese Economy: Analysis and Trade Implications," in Chalmers Johnson, Laura D'Andrea Tyson, and John Zysman, eds., *Politics and Pro-*

ductivity: How Japan's Development Strategy Works (New York: Ballinger, 1989). See also Ulrike Wassmann and Kozo Yamamura, "Do Japanese Firms Behave Differently? The Effects of Keiretsu in the United States," in Kozo Yamamura, ed., *Japanese Investment in the United States: Should We Be Concerned?* (Seattle: Society for Japanese Studies, 1989).

20. In late 1990, ICL passed into Fujitsu's hands largely as a consequence of its hardware dependence.

21. On this issue, see the remarks of Andy Grove, president of Intel Corporation, in *The New York Times*, May 2, 1990, p. D17. This problem is also the central motivation behind formation of the Computer Systems Policy Project, a new trade association and lobby of all of the top U.S.-owned computer producers.

22. In 1989, the Department of Defense reported that Japan had "significantly" overtaken the United States in six critical advanced technology areas: microelectronics circuit design and fabrication; preparation of compound semiconductors; machine intelligence and robotics; integrated optics; superconductivity; and biotechnology materials and processing. *Critical Technologies Plan for the Committees on Armed Services,* Department of Defense, May 5, 1989. Also, in March 1991, the Council on Competitiveness reported that the United States was weak, losing badly, or had lost one-third of the 94 generic technologies identified by the Council as critical to U.S. industrial competitiveness. *Gaining New Ground: Technology Priorities for America,* Council on Competitiveness, March 1991. The Department of Commerce recently made similar estimates for commercial technology. See U.S. Department of Commerce, *Emerging Technologies* (Washington, D.C.: GPO, 1990).

23. Fred Bergsten, "Exports: Rx for America?" *International Economic Insights,* Vol. 2, no. 1, January–February 1991.

24. *Ibid.*

25. *OECD Economic Outlook,* Vol. 47 (Paris: OECD, June, 1990).

26. Stephen Cohen, David J. Teece, Laura Tyson, and John Zysman, "Global Competition: The New Reality," *Working Paper of the President's Commission on Industrial Competitiveness,* Vol. 3, 1984. The definition of *competitiveness* given here is as follows:

> Competitiveness has different meanings for the firm and for the national economy. At the level of the individual firm, competitiveness is a fairly easy concept to define and understand. A firm's competitiveness is its ability to increase earnings by expanding sales and/or profit margins in the market in which it competes. This implies the ability to defend market position in the next round of competition, as products and production processes evolve. If the market is international in scope, so is the concept. Clearly, competitiveness is almost synonymous with a firm's long-run profit performance relative to its rivals.
>
> An analogue exists at the national level, but in our view it is much more complicated. A nation's competitiveness is the degree to which it can, under free and fair market conditions, produce goods and services that meet the test of international markets while simultaneously expanding the real incomes of its citizens. International competitiveness at the national level is based on superior productivity performance and the economy's ability to shift output to high productivity activities, which in turn can generate high levels of real wages. Competitiveness is associated with rising living standards, expanding employment opportunities, and the ability of a nation to maintain its international

obligations. It is not just a measure of the nation's ability to sell abroad, and to maintain a trade equilibrium. The very poorest countries in the world are often able to do that quite well. Rather, it is the nation's ability to stay ahead technologically and commercially in those commodities and services likely to constitute a larger share of world consumption and value-added in the future. Clearly, a nation's ability to compete internationally is reflected by its ability to maintain favorable terms of trade, which in turn governs the ease with which its citizens can maintain their international obligations while enjoying steadily rising real incomes.

National competitiveness will, of course, rest on competitive firms generating the productivity levels needed to support high wages, especially in growth sectors. The competitiveness of firms depends upon the quality and quantity of physical and human resources, the manner in which resources are managed, the supporting infrastructure of the economy, and the policies of the nation. National and corporate competitiveness are analytically distinct but practically intertwined. [p. 2]

27. Richard C. Marston, "Price Behavior in Japanese and U.S. Manufacturing," National Bureau of Economic Research, Working Paper No. 3364 (Cambridge, Mass.: National Bureau of Economic Research, 1990).

28. See, e.g., "Honda Prepares to Survive Yen Rise up to 120 to US Dollar," *Japan Economic Journal*, December 27, 1986, p. 1. The electronics data comes from confidential U.S. industry estimates.

29. William R. Cline, *United States External Adjustment and the World Economy* (Washington, D.C.: Institute for International Economics, 1989).

30. Paolo Guerrieri, *Technology and International Trade Performance of the Most Advanced Countries* (Rome: Department of Economics, University of Rome, 1990).

31. See, e.g., Robert Lawrence, *Can America Compete?* (Washington, D.C.: Brookings Institution, 1984).

32. This point is explicit with respect to machine tools in Bo Carlson and Stefan Jacobson, "What Makes the Automation Industry Strategic?" *Economics of Innovation and New Technology*, Vol. 1, no. 4 (1991), pp. 5–6; and for semiconductor manufacturing engineering and materials see Jay Stowsky, "The Weakest Link: Semiconductor Production Equipment, Linkages, and the Limits to International Trade," BRIE Working Paper No. 27, 1987; with respect to semiconductor and computer technology in Michael G. Borrus, *Competing for Control: America's Stake in Microelectronics* (Cambridge, Mass.: Ballinger, 1988); and in general MIT Commission on Industrial Productivity, The Working Papers of the MIT Commission on Industrial Productivity (Cambridge, Mass.: MIT Press, 1989).

33. For a skeptical view, see Jagdish Bhagwati, *Protectionism* (Cambridge, Mass.: MIT Press, 1988).

34. In the early 1980s, many analysts asserted that the share of manufacturing in American GDP had not declined. (See again, e.g., Lawrence, *Can America Compete?*) They argued that particular cases of industrial trouble were simply isolated though visible instances that did not describe the broad patterns in the American economy. A close look suggested that the numbers themselves were suspect. See Larry Mischel, *Manufacturing Numbers* (Washington, D.C.: Economic Policy Institute, 1988); and U.S. Congress, Office of Technology Assessment, *Paying the Bill*. A recent revision admits that real declines in manufacturing share of GDP occurred until the mid to late 1980s. (U.S. Department of Commerce, "U.S. Manufactures Trade Performance 1986 through

1990 Fourth Quarter" [Washington, D.C.: GPO, 1991]). Then, with the export boom and substantial direct foreign investment in U.S. industry, it appears that the manufacturing share of U.S. GDP began to rise again, reaching the previous peaks of the 1970s. Should we therefore be unconcerned? First, recall that this boom has been accomplished with falling wages and a hugely devalued dollar. As a consequence, employment and output in industry may grow, but the composition of production may shift: The domestic economy may slowly be ceding the market segments and core competences on which future competitive advantage—and strong future growth opportunities—lie. The sectoral evidence suggests that American manufacturing surged in lower value-added segments of particular industries; that apart from a few stellar performers like Boeing, America is becoming a low-wage, low-technology manufacturing economy. Paul Krugman confirms this in *The Age of Diminished Expectations: U.S. Economic Policy in the 1990s* (Cambridge, Mass.: MIT Press, 1990). Some also sought reassurance in data suggesting that the American problem was not its companies, which were believed to be globally competitive, but its weakness as a production location. See R. E. Lipsey and I. B. Kravis, *Productivity and Trade Share* (Washington, D.C.: National Bureau of Economic Research, March, 1984); R. E. Lipsey and I. B. Kravis, *Banca Nazionale del Lavoro Quarterly Review*, Vol. 153, No. 127 (June 1985). In fact, however, the data demonstrated something else—that U.S. firms were able, temporarily, to hold market position by moving production offshore to cheaper labor locations. There were, in fact, hints of serious problems later to come in sectors such as automobiles and electronics as firms delayed introducing innovative production technologies. The moves offshore often proved to be signs of weakness that generated greater weakness. And in any case, the competitiveness of the United States as a production location is of direct relevance to the security debate, as we elaborate later.

35. Jacques de Bandt, "Les Systemes Industriels: Dynamique, Croissance, Performances," *Revue d'économie industrielle*, no. 53, third trimester 1990.

36. Cohen and Zysman, *Manufacturing Matters*.

37. See the data and arguments in Nye, *Bound to Lead*.

38. Michel Albert, *Capitalisme Contre Capitalisme* (Paris: Seuil, 1991).

39. We identify two countries only, since to introduce more would presume substantial collective bargaining.

40. See OECD: Department of Economics and Statistics, *Quarterly National Accounts*, No. 4 (Paris: OECD, 1984).

41. Michael Dertouzos, Lester, and Solow, *Made in America*. This report of the MIT Commission on Industrial Productivity develops many of the same arguments on manufacturing changes advanced in this section. Many of the same arguments were developed in Cohen and Zysman, *Manufacturing Matters*, in 1987. *Made in America* continues an argument that was developing in the 1980s and is noteworthy for its excellent empirical material.

42. Policies favoring consumption and demand management are necessary to mass production because they increase the likelihood that production will be at near-full capacity—the only way that mass producers are profitable. This point is made explicitly, with respect to Keynesean policies, by Michael Piore and Chuck Sabel, *The Second Industrial Divide* (New York: Basic Books, 1984).

43. There is a debate over how to account for such developments. Two lines of analysis are suggestive. One, beginning with Alfred Chandler, emphasizes strategic choices and organizational innovation. See, in particular, Chandler's *The Visible Hand: The Managerial Revolution in American Business* (Cambridge, Mass.: Belknap Press, 1977). A second line is that of Oliver E. Williamson, building on the work on transac-

tions costs of Ronald Coase. See, e.g., Williamson's *Markets and Hierarchies, Analysis and Antitrust Implications: A Study in the Economics of Internal Organization* (New York: Free Press, 1975). In our view, Williamson's approach is an awkward means of addressing historical developments. It appears to be useful, however, in reconciling economists to historical and sociological reality.

44. For a fuller version of this argument, see Laura D'Andrea Tyson and John Zysman, "Developmental Strategy and Production Innovation in Japan," in Johnson, Tyson, and Zysman, eds., *Politics and Productivity*.

45. In the most recent detailed analysis of Japanese production innovations in the automotive industry, the authors explicitly reach an identical conclusion based on their findings. See James Womack, Daniel T. Jones, and Daniel Roos, *The Machine That Changed the World* (New York: Rawson Associates, 1990).

46. See "Manufacturing Innovation and American Industrial Competitiveness," *supra*.

47. See, e.g., Roos, Womack, and Jones, *The Machine That Changed the World*; Robert H. Hayes, Steven C. Wheelwright, and Kim B. Clark, *Dynamic Manufacturing: Creating the Learning Organization* (New York: Free Press, 1988); Ramchandran Jaikumar, "From Filing and Fitting to Flexible Manufacturing: A Study in the Evolution of Process Control," Working Paper (Cambridge, Mass.: Harvard Business School, 1988); and Peter Drucker, "The Emerging Theory of Manufacturing," *Harvard Business Review*, May/June 1990, pp. 94–102.

48. Piore and Sabel, *The Second Industrial Divide*, first elaborated this model emphasizing the Italian variants. For a German version, see Gary Herrigel, "Industrial Order and the Politics of Industrial Change: Mechanical Engineering," in Peter Katzenstein, ed., *Industry and Politics in West Germany* (Ithaca, N.Y.: Cornell University Press, 1989). Industrialists such as Romano Prodi, the former head of the Italian state holding company IRI, express concerns about the viability of this latter model. As a production system, we believe it is far less significant than the largely Japanese model described in this section.

49. There is a broad literature from which we derive what follows. See, e.g., from Europe, Benjamin Coriat, *L'Atelier et la Robot* (Paris: Christian Bourgeois, 1990); from the United States, Hayes, Wheelwright, and Clark, *Dynamic Manufacturing*, and Roos, Jones, and Womack, *The Machine That Changed the World*; from Japan, Yasuhiro Monden, *The Toyota Production System* (Norcross, Ga.: Industrial Engineering and Management Press, Institute of Industrial Engineers, 1983); and from the business press, e.g., Peter Drucker, "The Emerging Theory of Manufacturing."

50. The term *lean production* and this description come from Roos, Jones, and Womack, *The Machine That Changed the World*, who attribute it to Krafcik.

51. Rapid, cheap changeover is accomplished in a multitude of clever, relatively modest ways that include simple jigs, easy recalibration, and innovations in tool transport that increase mobility.

52. In semiconductor manufacturing, for example, such sources include materials purity, machinery precision, and environmental cleanliness.

53. On the microeconomics of this, see Coriat, *L'Atelier et la Robot*.

54. This characterization comes from a lecture by IBM's director of technology, James McGroddy, at Cornell University's Graduate School of Engineering, Distinguished Lecturer Series, May 1, 1989.

55. See, e.g., the discussion in Patrizio Bianchi, *Industrial Reorganization and Structural Change in the Automobile Industry* (Bologna, Italy: Collona di Economia Appli-

cata Editrice, 1989). Bianchi cites Stiglitz to remind us that vertical integration is a characteristic of mature industries that is open to change as those industries change. On the possibilities for such "de-maturity," see William J. Abernathy, Kim Clark, and Alan Kantrow, *Industrial Renaissance* (New York: Basic Books, 1983).

56. On the application of information networks in both manufacturing and services, see Francois Bar, Michael Borrus, and Benjamin Coriat, "Issues for Government Policy and Corporate Strategy," and Bar and Borrus, "From Public Access to Private Connections: Network Strategies in US Telecommunications" in *Information Networks and Comparative Advantage*; Vols. 1-3, OECD–BRIE Project (Paris: OECD, 1989), and the entire range of studies done under the auspices of the OECD–BRIE Telecommunications User Group Study. For an excellent appraisal of the "productivity paradox" that puts the problem into historical perspective, see Paul David, "Computer and Dynamo: The Modern Productivity Paradox in Historical Perspective," Working Paper No. 172 (Stanford, Calif.: Center for Economic Policy Research, 1989).

57. Jaikumar, "From Filing and Fitting to Flexible Manufacturing."

58. Cohen and Zysman, "Business, Economics, and the Oval Office."

59. Based on extensive industry discussion.

60. The exploration of how new world views emerge in labor and management is the strength of two quite different interpretations of business history. See Charles Sabel, *Work and Politics: The Division of Labor in Industry* (Cambridge [Cambridgeshire]; New York: Cambridge University Press, 1982), and Alfred Chandler, *Structure and Strategy: Chapters in the History of the Industrial Enterprise* (Cambridge, Mass.: MIT Press, 1962).

61. This claim is based on a sample of conversations with senior domestic and foreign managers that is not statistically significant, as well as on case studies of suppliers in the automobile and electronics industries done at BRIE. The point is widely acknowledged in industry and lies behind the formation of such collective entities as the National Center for Manufacturing Sciences. It is also supported by Womack, Jones, and Roos, *The Machine That Changed the World,* and by the detailed case studies done by the MIT Commission on Industrial Productivity. See Michael Dertouzos, Lester, and Solow, *Made in America.*

62. Ongoing work at BRIE has begun to document this for electronics. Broader support for the proposition can be found generally in the MIT study, Dertouzos et al., op. cit.

63. George N. Hatsopoulos, Paul R. Krugman, and Lawrence H. Summers, "U.S. Competitiveness: Beyond the Trade Deficit," *Science,* July 15, 1988.

64. As reported in Electronic News, May 21, 1990.

65. See Roos, Womack, and Jones, *The Machine That Changed the World.*

66. Confidential communication.

67. Moran, "The Globalization of America's Defense Industries," suggests a similar line of argument, though again, we disagree with his interpretation of the changing structure of the international economy.

68. This notion of the supply base is drawn from the work of Michael Borrus, from which we derive the definition and concerns set out here. For a very brief preliminary statement, see, e.g., Borrus, "Re-organizing Asia: Japan's New Development Trajectory and the Regional Division of Labor," BRIE Working Paper No. 53 (1992).

69. The economics of infrastructure is quite underdeveloped. In general, infrastructure is defined as being outside any individual firm, and as ubiquitously available, indivisible, and generating broad externalities (social gains that are not fully capturable by private firms). By this definition, our supply base notion (especially given technological

spillovers in advanced sectors such as electronics) qualifies as an infrastructure with the caveat that the open question remains of precisely how nationally "indivisible" it is. This is, of course, precisely the issue we examine in the text.

70. For example, Siemens executives confirm that the danger of dependence on Japanese competitors was one of the explicit rationales for Siemens' expensive move into memory chip production in the early 1980s.

71. As we suggest in a moment, the very largest firms are often able to obtain technology wherever it emerges and from whoever develops it; the small and medium-sized firms are much more dependent on national channels of technology flow.

72. Contrary to the arguments of analysts such as Bhagwati and Lawrence, industries are linked in more than simple input–output relations of no special significance. See, e.g., Bhagwati, *Protectionism,* and Lawrence, *Can America Compete?* Our work, like that of Cohen, Dosi, Freeman, Nelson, and Rosenberg (among many others), suggests that the composition of domestic production and the specific character of links among firms and sectors matter to technological and industrial development. See, e.g., Cohen and Zysman, *Manufacturing Matters;* Giovanni Dosi, Christopher Freeman, Richard R. Nelson, Gerald Silverberg, and Luc Soete, eds., *Technical Change and Economic Theory* (London and New York: Columbia University Press, 1988). In our view, recognition of this fact has been a central component of Japan's remarkable postwar economic transformation.

73. Dosi, Freeman, Nelson, Silverberg, Soete, op. cit. See Christopher Freeman, *Technology Policy and Economic Performance: Lessons from Japan* (London: Pinter, 1987), for the analysis applied to Japan.

74. On the concept of path dependence, see W. Brian Arthur, "Competing Technologies and Lock-In by Historical Events: The Dynamics of Allocation under Increasing Returns," Publication No. 43 (Stanford, Calif.: Stanford University, Center for Economic Policy Research, January 1985). Traditional studies usually attribute a large share of economic growth to technological development, but treat the technology component as residual. See, e.g., Edward Denison, *Why Growth Rates Differ: Postwar Experience in Nine Western Countries* (Washington D.C.: Brookings Institution, 1967). Until recently, most formal economic theory has treated technology either as invariant or as exogenous to economic analysis—that is, as independent of the existing allocation of resources. Such assumptions are not justifiable in modern industrialized economies. As a result, traditional models provide little help in understanding the long-term evolution of either technology or the economy.

75. The cyclical-versus-ladder approaches to scientific advance are described very well by Ralph E. Gomory, "Turning Ideas into Products," *The Bridge: Official Journal of the National Academy of Engineering,* Vol. 18, No. 1, Spring 1988.

76. Theodore H. Moran, "International Economics and National Security," *International Security,* Vol. 69, no. 5, Winter 1990/91.

77. Testimony by Stephen Cohen, Director, BRIE, UC Berkeley, to the Joint Economic Committee of the U.S. Congress, Sept. 5, 1990.

78. Stephen S. Cohen, presentation to the Joint Economic Committee of the U.S. Congress, September 23, 1987.

79. This is not to deny that R&D funds were often spent on activities that failed to produce winners. For example, the military funded many different efforts to build miniaturized electronic components that were unable to compete with the eventual winner, the integrated circuit (IC)—in part because it was prudent to back several competing approaches in a situation of great technological uncertainty. Even the IC,

whose creators generally credit their civilian R&D effort for the relevant advances, benefited from know-how spilling over from the detailed defense efforts, from defense funding of graduate education and research in electronics, and from funding of prototype electronic systems that demonstrated the efficacy of the fledgling technology. See, in general, Borrus, *Competing for Control,* Chapter 4, and the numerous sources cited there.

80. See Jay Stowsky, "Beating Our Plowshares into Double-Edged Swords: The Impact of Pentagon Policies on the Commercialization of Advanced Technologies," Berkeley Roundtable on the International Economy, Working Paper No. 17 (April 1986).

81. For details, see the discussion in Dertouzos, Lester, and Solow, *Made in America.*

82. Steven Vogel, "Japanese High Technology, Politics, and Power," Berkeley Roundtable on the International Economy, Research Paper No. 2 (March 1989), particularly Chapter 1.

83. See Dataquest Incorporated and Quick, Finan, and Associates, *The Drive for Dominance: Strategic Options for Japan's Semiconductor Industry* (San Jose, Calif.: Dataquest, 1988), pp. 4–7, citing Electronics Industry Association of Japan (EIAJ) data.

84. The examples that follow are culled from conversations with industry executives in military and nonmilitary electronics in the United States, Japan, and Europe.

85. McGroddy, Cornell's Distinguished Lecturer Series.

86. It is a well-accepted fact that military product development cycles are gruesomely long, usually resulting in military systems' incorporating electronic components that are several generations behind the existing state of the art. For example, it took eleven years for products incorporating the military's first very high speed integrated circuits (VHSICs) to appear on the market even though the VHSIC program's major purpose was *rapid* insertion of advanced components in weaponry.

87. Gomory, "Turning Ideas into Products," p. 12.

88. Sandholtz and his colleagues have phrased the problem nicely:

> In the short term, however, states frequently have to choose between them [power and wealth]. Advanced and civilian technologies increasingly converge on the same key sectors. The overlap encompasses the whole technological base of the nation: its research institutions, its industries, its pool of scientific and technical personnel, R&D investment pattern, in short the entire framework supporting innovation. The dilemma is that military objectives for technological development may shape the technical base in ways that handicap the corresponding civilian industries, or that commercial R&D will not produce the applications.

Wayne Sandholtz, Jay Stowsky, and Steve Vogel, "The Dilemmas of Technological Competition in Comparative Perspective: Is It Guns vs. Butter?" (Paper prepared for MacArthur Arms Control Seminar, Berkeley, Calif., Fall 1988), p. 1.

89. More generally, dominant organizations that control a critical resource shape the structure of subordinate organizations that require that resource. The subordinate organization adapts itself to obtain the critical resource. Those adaptations generally involve mimicking the structure of the dominant organization in order to provide better communication with those who make decisions. See John Zysman, *Political Strategies for Industrial Order: State, Market and Industry in France* (Berkeley: University of California Press, 1977).

90. See Kenneth N. Waltz, *The Theory of International Politics* (Reading, Mass.: Addison-Wesley, 1979).

91. "Globalization and Production." (Paper prepared for the BRIE Globalization and Production Conference, Berkeley, Calif., April 1991.)

92. Needless to say, this argument is widely disputed. For the skeptical view, see Gerald Segal, *Rethinking the Pacific* (Oxford: Clarenden Press, 1990), and Jeffrey Schott, "Is the World Devolving into Regional Trading Blocs?" (Washington, D.C.: Institute for International Economics, 1989). For a view parallel to ours, see Lawrence B. Krause, "Trade Policy in the 1990s: Goodbye Bipolarity, Hello Regions," *World Today*, Vol. 46, no. 5 (Royal Institute of International Affairs, May 1990).

93. These figures are based on data in Bureau d'Information et Prévision Economique, *Europe in 1992* (Paris: BIPE, October 1987).

94. This has been widely reported, and we elaborate it below. See, in general, the discussion in Japan Economic Institute, "Economic Regionalism: An Emerging Challenge to the International System," *JEI Report*, no. 25A, June 29, 1990, especially at pp. 5–6.

95. "Foreign Direct Investment in East Asia," prepared by Sylvia Ostrey for a joint research project designed by Sylvia Ostrey, Center for International Studies, University of Toronto, and the Berkeley Roundtable on the International Economy, 1992.

96. *Ibid.*

97. *Ibid.*

98. Here we use *sensitivity* to mean increased contact and exchange, *vulnerability* to mean sensitivities that are difficult to reverse, and *critical vulnerabilities* to mean those vulnerabilities that can threaten the stability of the political regime or the economy. See John Zysman, "The French State in the International Economy," in Katzenstein, ed., *Between Power and Plenty*. This usage closely parallels Waltz, *The Theory of International Politics*. For an example of critical vulnerability, see the discussion of the role of food in the creation of the nation state. Charles Tilly, ed., *The Formation of National States in Western Europe* (Princeton, N.J.: Princeton University Press, 1975).

99. Shintaro Ishihara, *The Japan That Can Say No: The New U.S.–Japan Relations Card* (New York: Simon & Schuster, 1991). Originally released as Shintaro Ishihara and Akio Morita, "The Japan That Can Say No: The New U.S.–Japan Relations Card," translated by the U.S. Department of Defense. With less controversy and greater technical accuracy, Japanese journalists, industrialists, and government officials have expressed similar views.

100. Figures from a presentation by the Keidanren's Kazuo Nukazawa in a presentation in the United Kingdom at the Royal Institute of International Affairs, July 27, 1990.

101. "The Rising Tide: Japan in Asia," special supplement, *Japan Economic Journal*, 1990, p. 4.

102. *Ibid.* See also Takashi Inoguchi, "Shaping and Sharing Pacific Dynamism," *The Annals of the American Academy of Political and Social Science*, Vol. 505, September 1989.

103. Data calculated from various sources by Lawrence Krause in "Pacific Economic Regionalism and the United States." (Paper prepared for the Symposium on *Impact of Recent Economic Developments on U.S.–Korea Relations and the Pacific Basin*, University of California–San Diego, November 9–10, 1990.)

104. There has been a considerable rise in the manufactured products share of total Japanese imports, from 31% in 1985 to about 50% in 1989. As the following table suggests, Japan has increased its imported manufactures from all major global sources, but the United States has lost relative position while Europe and Asian economies have gained, with Asian NICs gaining fastest.

Japan's Imports of Manufactured Products by Major Supplier
1985–89 (millions of dollars, c.i.f.)

	Total	U.S.	EC	Asian NICs
1985	40,157	14,243	7,691	5,689
1986	52,781	17,645	11,956	7,803
1987	65,961	17,672	15,146	12,456
1988	91,838	23,540	20,770	18,234
1989	106,111	28,119	24,193	20,495

Source: Japanese Ministry of Finance, Japan Economic Institute.

105. Yung Chul Park and Won Am Park, "Changing Japanese Trade Patterns and the East Asian NICs." (Paper prepared for National Bureau for Economic Research (NBER) Conference, October 19–20, 1992.)

106. Intra Asian data from Export–Import Bank of Tokyo Annual Report 1990.

107. *Ibid.*

108. Malaysian data from Malaysian Industrial Development Authority as cited in "The Rising Tide," *supra,* p. 4.

109. See the discussion and data in Park and Park, "Changing Japanese Trade Patterns."

110. MITI, 1987 White Paper on International Trade and Investment, (Tokyo: MITI, 1987), as cited in Japan Economic Institute, "Economic Regionalism."

111. Carl Goldstein, "Steering Committee: Japanese Car Makers Forge ASEAN Component Links," *Far Eastern Economic Review,* Vol. 147, No. 7, February 15, 1990, p. 67.

112. The argument here is fully set out and supported by Laura D'Andrea Tyson and John Zysman, "Developmental Strategy and Production Innovation," in Chalmers Johnson et al., eds., *Politics and Productivity.*

113. Bela Balassa and Robert Lawrence agree that Japan's trade structure differs from that of the other advanced countries. Saxonhouse applies Hechsher Olin models to reach different conclusions.

114. See Mordechai E. Kreinen, "How Closed is Japan's Market?" *World Economy,* Vol. 7, December 1988, pp. 529–541. On the MITI surveys, see the discussion in K. Iwata, "Changes of Economic and Trade Structure in the Pacific Basin Area." (Paper presented to International Economic Structure Research Group (group 3) Interim Report, Foundation for Advanced Information and Research (FAIR), Tokyo, Japan, June 1989.)

115. *Ibid.*

116. These percentages are from Edward Graham and Paul Krugman, *Foreign Direct Investment in the United States* (Washington, D.C.: Institute for International Economics, 1989).

117. This happened, for example, with Fuji-Xerox and TI-Japan.

118. This perspective is drawn from Wayne Sandholtz and John Zysman, "1992: Recasting the European Bargain," *World Politics,* Vol. 42, no.1 (October 1989).

119. Colette Herzog, "Les Trois Europes," in *Europe*, Economie Prospective Internationale Revue du CEPII Numéro Spécial (La Documentation Française 43, 3ème trimestre 1989).

120. *Ibid.*

121. The figures vary in the range given depending upon whose data set is used.

Our figures were obtained from national account statistics: main aggregates and detailed tables, 1987 (Geneva: United Nations, 1988). The comparisons were calculated at BRIE. Note that our figures differ significantly from those calculated by Michael C. Webb and Stephen D. Krasner, "Hegemonic Stability Theory: An Empirical Assessment," *Review of International Studies*, Vol. 15, no. 2, April 1989. We do not understand how they derived their numbers.

122. Communication between MM. Pandolfi de Bangemann and M. le President, M. Andriessen de Sir Leon Brittan, "L'Industrie Européenne de L'Electronique et de L'Informatique," Commission des Communautés Européennes, 1991.

123. See Steven Vogel's chapter, "The Power behind 'Spin-Ons': The Military Implications of Japan's Commercial Technology," in this book.

124. Yakushiji Taizo, "The Dynamics of Techno-Industrial Emulation: An Essay on the Growth Patterns of Industrial Pre-eminence and U.S.–Japanese Conflicts in High Technology," Berkeley Roundtable on the International Economy, Working Paper No. 15, Summer 1985.

125. Vogel, "Japanese High Technology," p. 46.

126. Conference, Royal Institute of International Affairs, Chatham House Council on Foreign Relations, London, December 1991.

127. Again we refer to the definitions of power elaborated in Caporaso and Haggard, "Power in International Political Economy," and Spiro and Caporaso, "Why Honest Theorists Disagree."

128. See the following two titles by Paul Krugman and Elhanan Helpman: *Trade Policy and Market Structure* (Cambridge, Mass.: MIT Press, 1989), and *Market Structure and Foreign Trade: Increasing Returns, Imperfect Competition, and the International Economy* (Cambridge, Mass.: MIT Press, 1985).

129. See, e.g., the work of Stephen Mark Haggard, *Pathways from the Periphery: The Politics of Growth in the Newly Industrializing Countries* (Ithaca, N.Y.: Cornell University Press, 1990).

130. See, e.g., Richard Rosencrance, *Rise of the Trading State: Commerce and Conquest in the Modern World* (New York: Basic Books, 1985).

131. Albert O. Hirschman's exploration of how nations can exploit the dependence of clients to capture gains from trade is the counterpoint to Rosencrance's notion of the trading state. See Hirschman's *National Power and the Structure of Foreign Trade* (Berkeley: University of California Press, 1945).

132. The entire issue of *International Security* (Summer 1989) is rooted in the old reality. The basic debate—the character of containment and the place of the third world—is an extension of the postwar argument. How, if our resources are now more limited, ought we to proceed? An excellent question, and we certainly favor the notion of finite containment.

Chapter 2

1. *Nihon Keizai Shimbun* (November 5, 1988).

2. According to DOD sources, these two projects are designed primarily to enable the SDIO to test experimental systems with Japanese components that are significantly more advanced than their American counterparts. In this way, the SDIO is able to "look into the future" and better anticipate what components they will be able to incorporate into future systems.

3. On the virtues of co-development, see Steven K. Vogel, "Let's Make a Deal: The U.S.–Japan Co-Technology Sphere," *The New Republic* (June 19, 1989), pp. 14–16.

4. Economic Planning Agency, Planning Department, *Nihon no sogo kokuryoku* [Japan's Comprehensive National Strength] (Tokyo: EPA, 1987).

5. John W. Dower refers to Japan as a major military "actor" in his "Japan's New Military Edge," *The Nation* (July 3, 1989), pp. 1, 18–22.

6. See Laura D'Andrea Tyson and John Zysman, "Developmental Strategy and Production Innovation in Japan," in Chalmers Johnson, Laura D'Andrea Tyson, and John Zysman, eds., *Politics and Productivity: How Japan's Development Strategy Works* (New York: Ballinger, 1989), pp. 59–140.

7. Masanori Moritani, *Gijutsu kaihatsu no showa-shi* [A Showa History of Technology Development] (Tokyo: Toyo Keizai, 1986), pp. 126–30. See Jeffrey A. Hart, "The Consumer Electronics Industry in the United States: Its Decline and Future Revival," in Francois Bar, Michael Borrus, Sabina Dietrich, Jeffrey Hart, and Jay Stowsky, *The U.S. Electronic Industry Complex*, Report to the U.S. Congress, Office of Technology Assessment (October 1988), for an overview of the decline of the U.S. consumer electronics industry; and see James E. Millstein, "Decline in an Expanding Industry: Japanese Competition in Color Television," in John Zysman and Laura Tyson, eds., *American Industry in International Competition: Government Policies and Corporate Strategies* (Ithaca, NY: Cornell University Press, 1983), pp. 106–41, for an analysis that focuses on Japanese firms' early conversion to all solid-state technology (1971 for most Japanese producers versus 1973–74 for RCA and Zenith).

8. Ampex, another U.S. firm, came out with the first videotape recorder in 1956. See Moritani, *Gijutsu kaihatsu no showa-shi*, pp. 145–50, and Richard S. Rosenbloom and Michael A. Cusumano, "Technological Pioneering and Competitive Advantage: the Birth of the VCR Industry," *California Management Review* (Summer 1987).

9. Moritani, *Gijutsu kaihatsu no showa-shi*, pp. 204–10.

10. Interview with Masanori Moritani, author and technology expert (Tokyo, July 20, 1987).

11. Thomas R. Howell, William A. Noellert, Janet H. MacLaughlin, and Alan Wm. Wolff, *The Microelectronics Race: The Impact of Government Policy on International Competition* (Boulder, Colo.: Westview, 1988), p. 217.

12. *Ibid.*, p. 56.

13. National Materials Advisory Board, Panel on Materials Science, *Advanced Processing of Electronic Materials in the United States and Japan* (Washington, D.C.: National Defense University Press, 1987), pp. 1–2. Also see Jay S. Stowsky, "Weak Links, Strong Bonds: U.S.–Japanese Competition in Semiconductor Production Equipment," in Johnson, Tyson, and Zysman, eds., *Politics and Productivity*, pp. 241–74.

14. U.S. Department of Defense, Office of the Undersecretary of Defense for Acquisition, *Report of the Defense Science Board Task Force on Defense Semiconductor Dependency* (February 1987).

15. On United States–Japan competition in semiconductors, see Michael Borrus, *Competing for Control: America's Stake in Microelectronics* (Cambridge, Mass.: Ballinger, 1988).

16. Ministry of International Trade and Industry, *Trends and Future Tasks in Industrial Technology: Summary of the White Paper on Industrial Technology* (Tokyo: MITI, 1988).

17. Science and Technology Agency, *Kagaku gijutsu hakusho* [Science and Technology White Paper] (Tokyo: STA, 1991), p. 375.

18. Moritani interview (July 20, 1987).

19. Science and Technology Agency, *Kagaku gijutsu hakusho*, p. 205.

20. *The New York Times* (March 7, 1988).

21. MITI data.

22. Science and Technology Agency, *Kagaku gijutsu hakusho*, pp. 318–21.

23. See Richard J. Samuels, *Research Collaboration in Japan*, M.I.T.–Japan Science and Technology Program Working Paper (February 1987). A number of authors have suggested that the impact of Japan's cooperative research projects is often overrated. See, for example, George R. Heaton, Jr., "The Truth about Japan's Cooperative R&D," *Issues in Science and Technology* (Fall 1988).

24. As cited in Jacques S. Gansler, "The Need and Opportunity for Greater Integration of Defense and Civilian Technologies in the United States" (unpublished paper, 1987), p. 13.

25. For a discussion of the Japanese perspective on "spin-ons," see Richard J. Samuels and Benjamin C. Whipple, "Defense Production and Industrial Development: The Case of Japanese Aircraft," in Johnson, Tyson, and Zysman, eds., *Politics and Productivity*, pp. 275–318.

26. *Aviation Week and Space Technology* (March 19, 1990).

27. Bruce D. Nordwall, "Electronic Technology to Dominate Next Generation of Weapons," *Aviation Week and Space Technology* (June 6, 1988), pp. 81–85.

28. *Nikkei Business* (May 11, 1987), p. 15.

29. These will be discussed in more detail later in the chapter.

30. *Voice* (September 1987), p. 95.

31. U.S. Department of Defense (for the Committees on the Armed Services, U.S. Congress), *Critical Technologies Plan* (March 15, 1990), p. 11.

32. U.S. Department of Defense, Office of the Undersecretary of Defense for Acquisition, *Japanese Military Technology: Procedures for Transfers to the United States* (1986).

33. Interview with Gregg Rubinstein, director of plans and policy, international operations, Grumman International, Inc. (Washington D.C., June 3, 1987).

34. Interview with Jamieson C. Allen, director of research and development exchange, Mutual Defense Assistance Office (Tokyo, June 26, 1987); and *JEI Report*, no. 30A (August 7, 1987).

35. U.S. Department of Defense, Office of the Undersecretary of Defense for Research and Engineering, Defense Science Board Task Force Report, *Industry-to-Industry International Armaments Cooperation Phase II—Japan* (1984).

36. U.S. Department of Defense, Office of the Undersecretary of Defense (Acquisition) Research and Advanced Technology, *Electro-Optics and Millimeter Wave Technology in Japan: Final Report of the DOD Technology Team* (May 1987).

37. The U.S. government has published a number of reports stressing the benefits of gaining access to Japanese dual-use technology. These include U.S. Department of Defense, Office of the Undersecretary of Defense for Acquisition, Defense Science Board Report, *Defense Industrial Cooperation with Pacific Rim Nations* (October 1989); and U.S. Congress, Office of Technology Assessment, *Arming Our Allies: Cooperation and Competition in Defense Technology* (Washington, D.C.: GPO, 1990).

38. Allen interview (July 11, 1988).

39. DOD sources.

40. "Wagakuni no boei sobi—sono gijutsu to gyokai no doko" [Our Country's Defense Equipment—Technology and Trends in the Industry], Part 2 of a seven-part series, *Kokubo* (November 1986), p. 109.

41. Japan Defense Agency, *Boei hakusho 1991* [Defense White Paper 1991] (Tokyo: JDA, 1991), p. 305.

42. Adachi interview (Tokyo, July 14, 1988).

43. Komoda interview (Tokyo, July 14, 1988).

44. Interview with senior JDA official (July 1988).

45. *Ibid.*

46. Asahi Shimbun Economic News Desk, *Miriteku pawaa: kyukyoku no nichibei masatsu* [Militech Power: The Ultimate United States–Japan Friction] (Tokyo: Asahi, 1989), p. 70.

47. Ono interview (Tokyo, July 6, 1987).

48. John O'Connell, "Strategic Implications of the Japanese SSM-1 Cruise Missile," *Journal of Northeast Asian Studies* (Summer 1987), p. 54.

49. "Japan Uses SSM-1 Expertise to Develop Cruise Missile," *Aviation Week and Space Technology* (March 21, 1988), p. 59.

50. Interview with Hiroshi Tajima, deputy general manager, Guided Weapons Department, and Takeki Wani, deputy general manager, Planning Department, Aircraft and Special Vehicle Headquarters—both of Mitsubishi Heavy Industries, Ltd. (Tokyo, July 8, 1988).

51. Asahi Shimbun Economic News Desk, *Miriteku pawaa,* pp. 13–14. For a more in-depth look at Japanese missile development, see parts 1 (October 1986) and 2 (November 1986) of a seven-part series in *Kokubo*.

52. Interview with senior JDA official (July 1988).

53. *Nihon Keizai Shimbun* (January 4, 1991).

54. *Kokubo* (October 1986), p. 31.

55. Interview with Shigeru Aoe, director of the Space Planning Bureau, Research and Development Division, Science and Technology Agency (Tokyo, July 13, 1988).

56. Interview with James G. Beitchman, vice-president, Communications Satellite Corporation, Comsat Far East Operations (Tokyo, July 23, 1987).

57. *Science and Technology in Japan* (August 1987), pp. 18–19.

58. Japan signed the NPT in 1976.

59. Nakagawa interview (Tokyo, July 29, 1987).

60. See Leonard S. Spector, *Going Nuclear* (Cambridge, Mass.: Ballinger, 1987), on nuclear technology.

61. Cited in Malcolm McIntosh, *Japan Rearmed* (London: Frances Pinter, 1986), p. 58.

62. Komoda interview (July 14, 1988).

63. Interview with senior American executive (July 1987).

64. Interview with W. Stephen Piper, president of InTecTran, Inc. (Washington, D.C., May 31, 1988).

65. This argument is made at greater length in Chapter 1 of this book.

66. Japanese analysts have not refrained from speculating about the military potential of Japan's high technology. In *Haiteku boei no susume* [A Case for the High-Tech Defense of Japan] (Tokyo: Simul, 1985), Kaoru Murakami suggests that rather than trying to imitate the superpowers, Japan should exploit its technological strengths and defend itself with high-tech weaponry. In *Gunji robotto senso* [Military Robot War] (Tokyo: Diamond, 1982), Masahiro Miyazaki envisions a world in which Japan's advanced robots, rather than its citizens, fight the country's wars.

67. Robert J. Art, "The Influence of Foreign Policy on Seapower," in Robert J. Art and Kenneth N. Waltz, eds., *The Use of Force* (Lanham, Md.: University Press, 1983), p. 186.

68. The Japanese defense debate is discussed at greater length in Steven K. Vogel, *A New Direction in Japanese Defense Policy: Views from the Liberal Democratic Party Diet Members,* Occasional Papers/Reprints Series in Contemporary Asian Studies, University of Maryland School of Law (1984); and Mike M. Mochizuki, "Japan's Search for Strategy," *International Security* (Winter 1983–84).

69. *Nihon Keizai Shimbun* (April 27 and 29, 1980).

70. On United States–Japan defense cooperation, see Gregg Rubinstein, "U.S.–Japan Security Relations," *The Fletcher Forum* (Winter 1988); and Norman D. Levin, *Japan's Changing Defense Posture,* RAND Note (June 1988).

71. The *Boei hakusho 1991* [Defense White Paper 1991] states that defense production accounted for 0.54 percent of total production in 1989, up from 0.36 percent in 1980 (p. 305).

72. Saito interview (Tokyo, August 3, 1987).

73. *Nikkei Business* (May 11, 1987), p. 13.

74. Samuels and Whipple, "Defense Production and Industrial Development," use the metaphor of a tree to explain why some Japanese planners feel that the aerospace industry is so important to technological development: The aerospace industry is a stem that is connected to both the "roots" (underlying technologies) and the "fruits" (related industries) of the tree. The point is not so much that one part of the tree is more important than another, but that the parts depend on each other for their own healthy development.

75. Komoda interview (July 14, 1988).

76. *Asahi Shimbun* (September 17, 1988).

77. *Chuo Koron* (July 1989), pp. 184–98.

78. Interview with senior JDA official (July 1988).

79. In the Lower House, the LDP won 286 seats, compared to 141 for the JSP, 46 for Komeito, 16 for the JCP, 14 for the DSP, 5 for others, and 4 for independents.

80. See J.A.A. Stockwin, *Japan: Divided Politics in a Growth Economy,* 2nd ed. (London: Norton, 1982), pp. 89–91.

81. Interview with Kunio Kinjo, senior staff, Public Relations Committee, Liberal Democratic Party (Tokyo, June 23, 1987); and interview with Masakatsu Shinkai, director, Communications Division, Equipment Bureau, Japan Defense Agency (Tokyo, June 24, 1987).

82. *Asahi Shimbun* (May 29, 1990).

83. Kato interview (Tokyo, July 12, 1988).

84. *Japan Times* (December 20, 1990).

85. Martin Libicki, Jack Nunn, and Bill Taylor, *U.S. Industrial Base Dependence/Vulnerability: Phase II—Analysis,* Report for the Mobilization Concepts Development Center, Institute for National Strategic Studies, National Defense University (November 1987), pp. 5–7.

86. U.S. Department of Defense, *Report of the Defense Science Board Task Force.*

87. U.S. Congress, Office of Technology Assessment, *The Defense Technology Base: Introduction and Overview* (Washington, D.C.: U.S. GPO, 1988), p. 40.

88. Interview with Martin C. Libicki, a professor at National Defense University (Tokyo, June 20, 1987).

89. *Sekai* (January 1988), p. 82.

90. Interview with senior JDA official (July 1988).

91. Allen interview (July 11, 1988); and *JEI Report,* no. 16B (April 22, 1988), p. 9.

92. Mullen interview (Washington, D.C., May 31, 1988).

93. Science and Technology Agency, *Kagaku gijutsu hakusho,* p. 318.

94. David B. H. Denoon, ed., *Constraints on Strategy: The Economics of Western Security* (Washington, D.C.: Pergamon-Brassey's, 1986), p. 208. Also see the chapter in Denoon's book on "Japan and South Korea" written by Walter Galenson and David W. Galenson, pp. 152–94.

95. Science and Technology Agency, *Kagaku gijutsu yoran* [Indicators of Science and Technology] (Tokyo: STA, 1990), pp. 150–53.

Chapter 3

1. The first part of this piece draws heavily from Wayne Sandholtz and John Zysman, "1992: Recasting the European Bargain," *World Politics,* Vol. 42, no. 1 (October 1989), pp. 95–128.

2. See Centre for Business Strategy, *1992 Myths and Realities* (London: London Business School, 1989).

3. Clemente Signoroni, "The ECU, a success Factor for the 1993 Community Market" (address given in Madrid, January 12, 1989), p. 6.

4. Enrique Baron, *Europe 92: Le Rapt du Futur* (Paris: Editions Bernard Coutas, 1989), p. 88.

5. We have no intention of providing a detailed history of the EEC; that story has been well told many times. We seek only to show that the major elements of that history fit the analytical framework we are proposing here.

6. Of course, many of the early students of European integration recognized that structural changes due to the war were crucial in triggering the process. See, for example, Leon N. Lindberg and Stuart A. Scheingold, *Europe's Would-Be Polity* (Englewood Cliffs, N.J.: Prentice-Hall, 1970), especially Chapter 1.

7. See Walter Yondorf, "Monnet and the Action Committee: The Formative Period of the European Communities," *International Organization* 19 (Autumn 1965).

8. The regional development funds had a precursor in development programs created at Italy's insistence in 1956. They acquired more importance after the accession of Britain and Ireland, and have become vital elements of the EC bargain since the addition of Greece, Spain, and Portugal.

9. Tommaso Padoa-Schioppa, *Efficiency, Stability and Equity* (Oxford: Oxford University Press, 1987).

10. Helen Wallace and Wolfgang Wessels, *Towards a New Partnership: The EC and EFTA in the Wider Western Europe,* Occasional Paper No. 28, European Free Trade Association, Economic Affairs Department (March 1989), p. 4.

11. Strategic games are useful heuristic devices that can help us reason about structured situations by clarifying the logic of interaction. In this instance, whatever the general methodological case, substantial investment in specifying and manipulating a multiplayer, multi-issue game will have limited payoffs in our understanding of 1992. Indeed, the crucial analytic issues must be resolved long before a set of games can even be devised. Games of strategic interaction require preference functions for each player. With 1992, decisionmakers do not possess the intellectual means to foresee alternative outcomes, much less rank them. Game theory, as even its most enthusiastic proponents recognize, cannot yet deal with changing preferences. Given all of these lacunae and uncertainties, game models of the international interactions involved in 1992 cannot possibly capture the political dynamics that matter. Behind the games are the crucial factors: political strategies, constraints, and leadership.

Nor is this really a problem for theories of collective action, as traditionally conceived in political science. The problem is not one of inducing actors to contribute to production of a collective good (i.e., avoiding free riders). The institutional structure of the community compels participation and shared leadership. At issue are the areas that should be opened to joint policymaking and the institutional arrangements that might prove acceptable to the parties. Not only are there substantial risks and costs for all, but imposing European decisions on domestic politics requires domestic political action by the national executive, not just acquiescence in the European Commission and Council of Ministers. There are, in other words, multiple layers of politics.

12. This is not a matter of elite learning that can be explained by theories of learning. Our proposition would clearly be that changed circumstances, not increased knowledge, have altered behavior. By knowledge we would mean formally specified relationships (information and theories about it) that suggest what outcomes will result from what causes. It is not a better understanding of an existing situation, but the discovery of a new situation, that is at issue. The necessary ingredient for adaptation is therefore vision and leadership, an image of arrangements or relationships that will respond to new tasks and the skill to mobilize diverse groups to construct that future. It is not greater technically rooted knowledge, but politically founded insight, that is called into play.

13. Some business and government leaders involved in 1992 are, in fact, trying to sidestep normal coalition politics in order to bring about domestic changes.

14. We have attempted (without fully succeeding) to distinguish between the politics of coalitions and the role of institutions in shaping the present response. In the first European movement that established the Coal and Steel Community, and then the EEC, there were no European institutions shaping and activating the players. Now there are, and the game is consequently quite different. The most important "spillover" probably lies in the creation of a permanent advocate of more extensive integration, and a location for such developments to occur.

15. Andrew Moravcsik, "Negotiating the Single European Act," *International Organization* (Winter 1981). Moravcsik seeks to distinguish his position from our view, presented in *World Politics,* by laying emphasis on the governmental deals leading to the Single European Act. Each article identifies and acknowledges the role of identical elements; each notes that the community is a bargain among governments; each notes the changing political context at the national level that created the basis in which governments could make new bargains. Except in emphasis, these tales do not differ, and each version can support the argument of the other author. Several points need clarifying. We find it confusing to be shoved into a neo-functionalist category because of our use of the word *elites.* Our argument explicitly contends that the neo-functionalist spillover process would not have produced a new bargain. It argues that the community actors were crucial in the recasting of the bargains and in the process of redefining conceptions. We place greater emphasis than does Moravcsik on Community institutions and business-sector involvement. Moravcsik focuses on the details of the negotiations; we try to capture the broader international and institutional context. He trains his analytic eye on the Single Act; we examine the larger movement called "1992."

16. The analytic histories of the new Europe are now beginning to appear. One of the best is David R. Cameron, "The 1992 Initiative: Causes and Consequences" in Alberta M. Sbragia, ed., *Euro-Politics: Institutions and Policymaking in the "New" European Community* (Washington, D.C.: The Brookings Institution, 1992).

17. Stephen Cohen, "Informed Bewilderment," in Stephen Cohen and Peter Gourevitch, eds., *France in a Troubled World Economy* (London: Butterworth, 1983). Cohen makes this point here in a particularly clear and jargon-free fashion.

18. See John Zysman, *Governments, Markets, and Growth: Finance and the Politics of Industrial Change* (Ithaca, N.Y.: Cornell University Press, 1983), Chap. 1.

19. There is both a European and an American school of discussion. In the United States, the debate is led by Charles Sabel and Michael Piore. Their book, *The Second Industrial Divide* (New York: Basic Books, 1984), brought many of the issues into the public arena, though the scholarly work underlying it is much more important. In Europe the group is diverse, including Robert Boyer, Benjamin Coriat, Giovanni Dosi, and Jacques Mistral. A particularly interesting version of the debate is found in Peter Hall, ed., *International Journal of Political Economy: European Labor in the 1980s*, Vol. 17, No. 3 (Fall 1987).

20. Robert Boyer makes this point particularly well in a diverse set of his papers.

21. There was a another twist, as well: David Flanagan at Stanford University has argued that companies hesitated to invest in training new workers and that, when they had to expand, tended to hire back those that had been laid off, creating a substantial pool of young unemployed.

22. There is a parallel story in Britain some fifteen years earlier. In the early 1960s, the British Labour Party had refused to devalue the pound sterling even when it meant effectively abandoning a growth strategy. The debate on sterling was suppressed within the party; the outcome was settled only when there was truly no choice left. See Stephen Blank, "Britain: The Politics of Foreign Economic Policy," in Peter J. Katzenstein, ed., *Between Power and Plenty: Foreign Economic Policies of Advanced Industrial States* (Madison: University of Wisconsin Press, 1978); Anthony Howard, ed., *The Crossman Diaries, Selection from the Diaries of a Cabinet Minister 1964–1970* (London: Methuen Paperbacks, 1979).

23. Hall, ed., *International Journal of Political Economy*.

24. Philip Revzin, "Italians Must Change Their Business Style in Integrated Europe," *Wall Street Journal* (November 21, 1988), p. 1. Italian businessmen quoted in the article express the same sentiment.

25. Michael Borrus et al., *Telecommunications Development in Comparative Perspective*, BRIE Working Paper No. 14 (Berkeley, Calif.: BRIE, 1985), p. 38.

26. This is a radical simplification, but it captures the essence of events. For more detail, see Wayne Sandholtz, *High-Tech Europe: The Politics of International Cooperation* (Berkeley: University of California Press, 1991).

27. Wallace and Wessels, *Towards a New Partnership*.

28. Michael Emerson, *1992 and After*. The White Paper proposals can be grouped into sets.

  a. Liberalization of government procurement; essentially opening national procurement to outside bidders.
  b. Technical norms; the largest number of proposals set technical standards that otherwise preclude movement of goods through Europe.
  c. Transport services.
  d. Agricultural border taxes and subsidies.
  e. National restrictions in the Community's external trade relations; these matters are not strictly an element of the "internal bargain" but are included here to be complete.
  f. Abolition of fiscal frontiers; there would be no need to assess taxes at the border.
  g. Financial services, including banking, stock markets, and related services and insurance; this set aims boldly at creating a European capital market.

29. Padoa-Schioppa, *Efficiency, Stability and Equity*; Emerson, *1992 and After*; Paolo Cecchini, *The European Challenge, 1992: The Benefits of a Single Market* (Aldershot: Gower, 1988).

30. "Europe's Internal Market," *The Economist* (July 9, 1988), pp. 6, 8.

31. This view is sometimes expressed in EEC materials lauding 1992. Not everyone would agree; they would cite budget initiatives in 1970 and 1975 and the direct election of the European Parliament. We would cite the Single European Act because it rejects the national veto.

32. "Europe's Internal Market," *supra,* Survey, p. 6.

33. *Europe 1990* (Brussels: Chez Philips S.A., no date), p. 5, emphasis in the original.

34. Rob van Tulder and Gerd Junne, *European Multinationals in Core Technologies* (New York: John Wiley & Sons, 1988), pp. 214–15.

35. Based on interviews and discussions.

36. Quoted in van Tulder and Junne, *European Mutinationals,* p. 215, fn. 8.

37. Axel Krause, "Many Groups Lobby on Implementation of Market Plan," *Europe* (July/August 1988), p. 24.

38. Another business group collaborating with the Commission and actively promoting the 1992 process is the Union of Industrial and Employers' Confederations in Europe (UNICE), which includes over thirty industrial associations throughout Europe. The secretary-general of UNICE, Zygmunt Tyszkiewicz, spoke of the group's working groups and lobbying: "Nine-tenths of our work comprises the regular, invisible interchange of ideas between our experts and the EC Commission's civil servants." Cited in *ibid.*

39. See Craig Forman, "European Firms Hope Swapping Stakes Gives Them 1992 Poison-Pill Protection," *Wall Street Journal* (October 18, 1988), p. A26.

40. Michael Porter, "Europe's Companies after 1992," *The Economist* (June 9, 1990), p. 18.

41. Michael Emerson, "1992 as Economic News" (unpublished essay, Brussels, November 1, 1988), p. 5.

42. Francois Perigot, "L'Europe ardente obligation," *Politique industrielle,* no. 10 (Winter 1988).

43. Phillipe Schmitter, "A Revised Theory of Regional Integration," *International Organization* 24 (Autumn 1970).

44. These conclusions are drawn from a map of global GDP developed by the BIPE (Bureau d'Informations et de Prévisions Economiques) and relabeled with their permission by BRIE. BIPE calculated these numbers from OECD figures. The map is in "La France dans l'Europe de 1993," Vol. 1, pp. 8 and 9 (no date). This information is taken both from BIPE, *Europe in 1992* (Paris: BIPE, October 1987) and from Gerard Lafay and Colette Herzog with Loukas Stemitsiotis and Deniz Unal, *Commerce international: La fin des avantages acquis* (Paris: Economica, 1989), with figures taken from diverse charts and tables in Section 1.

45. Lafay and Herzog et al., *Commerce international.*

46. Colette Herzog, "L'ouverture du marche europeen: Une autre vision," *Economie Prospective Internationale: Revue du CEPII* (first trimestre 1988), pp. 81–89.

47. Lafay and Herzog et al., *Commercial international,* from diverse charts and tables in Section 1.

48. The United States, at the moment of its economic and political dominance, structured the global trading system (as is so often repeated) around these principles and embedded them in its own legal practice. The reality, of course, has been more complicated. It has involved a series of exceptions to these principles. One set of these exceptions, made from a position of economic strength, was made for foreign policy objectives. The

United States—at least in its conception—opened its market to Japan, the developing economies, and Europe and tolerated trade discrimination. The second set of exceptions, made from a position of sectoral weakness, has involved bilateral bargains to contain imports in specific sectors: autos, textiles, steel. Those sectoral bargains were often made to accommodate those who would challenge the general principles of trade policy. Importantly, the level of imports that triggers substantial protectionism has been quite high. The delicate hypocrisy has been more difficult to maintain as the coalition in support of open trade has narrowed. The balance between principles of multilateralism and a reality of bilateralism has slowly tilted toward bilateralism as a growing number of sectors have achieved protection, and, in the years of the deficit, that tilt has become more pronounced.

49. For the difference between general and specific reciprocity, see Jagdish Bhagwati, *Protectionism* (Cambridge, Mass.: MIT Press, 1988).

50. This section is drawn from Michael Borrus and John Zysman, "Industrial Strength and Regional Response: Japan's Impact on European Integration" in Gregory F. Treverton, *The Shape of the New Europe* (Washington, D.C.: Council on Foreign Relations, 1992).

51. James Womack et al., *The Machine that Changed the World* (New York: Macmillan, 1990). This is a seminal work on the notion of "lean" production. They attribute the phrase to one of their colleagues, John Krafcik.

52. Laura Tyson and John Zysman, "Developmental Strategy and Production Innovation in Japan," in Chalmers Johnson, Laura Tyson, and John Zysman, eds., *Politics and Productivity: How Japan's Development Strategy Works* (New York: Ballinger, 1989).

53. Admiral Sir James Eberle, director, The Royal Institute of International Affairs. (Valedictory Address, December 12, 1990.)

54. The analysis that follows was undertaken with Costas Tsatsaronis. See Giovannini, *Limiting Exchange Rate Flexibility* (Cambridge, Mass.: MIT Press, 1989); F. Giavazzi, "The Exchange Rate Question in Europe," in Ralph C. Bryant, ed., *Macro Economic Policies in an Interdependent World* (Washington, D.C.: Brookings Institution, 1989); P. De Grauwe, "Is the EMS a DM Zone?" CEPR Discussion Paper, 1989; and Charles Wyplosz, "Asymmetry in the EMS Intentional or Systemic," *European Economic Review*, Vol. 33, No. 2/3 (March 1989), pp. 310–20.

55. B. Eichengreen, "One Money for Europe? Lessons from the U.S. Currency Union," *Economic Policy*, no. 9 (London, April 1990).

56. A. Casella, "Participation in a Currency Union" (CEPR Working Paper, 1990); and A. Casella and J. Feinstein, *Managing a Common Currency in a European Central Bank* (Cambridge, Mass.: Cambridge University Press, 1989).

57. See Wayne Sandholtz, "The EC and the East: Integration after the Revolutions of 1989," *Dialogue: A Magazine of International Affairs* 2 (Summer 1990), pp. 12–16.

58. Derived from Statistical Office of the European Communities, *Basic Statistics of the Community*, 27th ed. (Luxembourg: Office for Official Publications of the European Communities, 1990), pp. 270, 272.

59. Stage 1 includes participation in the EMS exchange rate mechanism by all Community currencies, removal of restrictions on private use of the ECU, a common financial market, enhanced economic and fiscal policy convergence, and increased monetary policy coordination.

60. David Buchan, "Mitterrand Urges Early Talks on EC Monetary Integration," *Financial Times* (October 26, 1989), p. 1.

61. Laura Raun, "Dutch Back Mitterrrand's Plan to Accelerate EMU," *Financial Times* (November 21, 1989), p. 2.

62. The final blow may have been Sir Geoffrey Howe's dramatic resignation from the Cabinet. In his speech to Parliament, Howe included a blistering criticism of Thatcher's stubborn opposition to increased European integration in monetary and political affairs.

63. The rate of its acceptance would not be a political choice, but European countries would use fewer dollars to intervene in exchange rate markets and there would be at least some demand for ECUs by third-country central banks for intervention purposes. Private-sector acceptance of the currency would turn on the ease of transactions. Crucially, to the extent that the ECU will be successful in replacing the dollar as a world currency, it will allow the EC to claim profits from international seignorage—the right to create money—by virtue of the fact that a European central bank will be able to issue debt-denominated assets of which it controls the supply. Such seignorage becomes a nonbudget means of providing a central authority with the equivalent of transfer payments to hold a monetary union together. European authorities are clearly aware of this and have begun to discuss the problems of seignorage in a reconstructed international monetary order in major Community documents.

64. William Walker and Philip Gummett, "Britain and the European Armaments Market," *International Affairs* (Summer 1989), pp. 435–42.

65. See Sandholtz, *High Tech Europe,* Chaps. 7, 8, and 9.

Chapter 4

1. Likewise, the Pentagon's recently announced plans to trim spending by merely prototyping future weapons systems (that is, building just a few of each design as a "technology demonstrator") fly in the face of current commercial practice here and abroad, which emphasizes design-for-manufacturability and iterative learning-by-doing from successive production runs. Instead of adapting U.S. military requirements to exploit product and process advances emerging from the commercial sector, Pentagon planners seem intent on creating isolated technological "arsenals" manned by monopoly producers of specialized military systems.

2. For a detailed account of the Army Signal Corps's involvement in the development of the transistor, see T. Misa, "Military Needs, Commercial Realities, and the Development of the Transistor, 1948–1958," in M. R. Smith, ed., *Military Enterprise and Technological Change: Perspectives on the American Experience* (Cambridge, Mass.: MIT Press, 1985).

3. R. Levin, "The Semiconductor Industry," in R. Nelson, ed., *Government and Technical Progress: A Cross-Industry Analysis* (New York: Pergamon Press, 1982), p. 58. Bell's continuing policy of swift public dissemination of its research results was influenced primarily by the antitrust suit that the U.S. Department of Justice initiated against AT&T in 1949. According to the terms of the 1956 consent decree that eventually ended the suit, Bell Labs continued to act as a sort of national research facility, disseminating basic solid-state technology and channeling the energies of commercial semiconductor firms to the search for broader applications.

4. Misa, "Military Needs," pp. 274–75.

5. For more on the notion of surge capability, see S. Melman, *Profits without Production* (New York: Alfred Knopf, 1983).

6. M. Borrus, *Competing for Control: America's Stake in Microelectronics* (Cambridge, Mass.: Ballinger, 1988), p. 66.

7. Bell Telephone Lab Report, 1957, cited in Misa, "Military Needs," p. 268. See also Borrus, *Competing for Control,* p. 61.

8. Military patronage of relatively new and innovative semiconductor producers, such as TI, grew in significance because most established commercial electronics producers were slow to recognize the revolutionary potential of solid-state technology. The early transistors were less reliable than vacuum tubes and more expensive; except for its adaptation to the manufacture of hearing aids, for which its compactness made it especially well suited, the transistor was not regarded as an economical substitute for vacuum tubes for most consumer electronics products. According to E. Braun and S. MacDonald, "Despite the early interest in the transistor as a better valve [tube], the transistor was so radically different from the valve in the way it worked, in the way it could be manufactured and sold, and in its apparent potential, that it could not be comfortably accommodated within the existing electronics industry without changes that that industry was then unwilling or unable to make." Braun and MacDonald, *Revolution in Miniature: The History and Impact of Semiconductor Electronics* (London: Cambridge University Press, 1978), p. 69.

9. W. R. Stevenson, "Miniaturization and Microminiaturization of Army Communications—Electronics, 1946–1964," U.S. Army Electronics Command, Fort Monmouth, New Jersey, Historical Monograph I (unpublished manuscript, 1966), pp. 125 and 130. Cited in Misa, "Military Needs," p. 282.

10. Misa, "Military Needs," pp. 283–84.

11. *Ibid.*, p. 285.

12. A year later, Bell Labs developed a similar "preferred-device" program to streamline transistor development for Bell System applications.

13. In fact, most of the early military contracts went not to innovative start-ups such as Transitron, Motorola, or Texas Instruments, but to established suppliers of soon-to-be-outmoded vacuum tubes, such as General Electric, Western Electric, Sylvania, Raytheon, and RCA. As late as 1959, the big firms were awarded 78 percent of the federal research money for learning how to manufacture cheaper, more reliable transistors, even though they accounted, at that time, for only 37 percent of the transistor market. Braun and MacDonald, *Revolution in Miniature,* p. 81, cited by R. DeGrasse in J. Tirman, ed., *The Militarization of High Technology* (Cambridge, Mass.: Ballinger, 1984), p. 91. Also mentioned in R. Reich, *The Next American Frontier* (New York: Times Books, 1983), pp. 190–91.

14. Cited in Levin, "The Semiconductor Industry," p. 61.

15. N. Asher and L. Strom, *The Role of the Department of Defense in the Development of Integrated Circuits* (Arlington Institute for Defense Analysis, 1977), pp. 4, 17. Also cited in M. Borrus, J. Millstein, and J. Zysman, *U.S.–Japanese Competition in the Semiconductor Industry* (Berkeley: Institute of International Studies, University of California, 1982), p. 17.

16. Asher and Strom, *The Role of the Department of Defense,* p. 17.

17. This point has been made by J. Utterback and A. Murray, *The Influence of Defense Procurement on the Development of the Civilian Electronics Industry* (Cambridge, Mass.: MIT Center for Policy Alternatives, 1977), p. 3; I. Magaziner and R. Reich, *Minding America's Business* (New York: Harcourt, Brace, Jovanovich, 1982); and Borrus, *Competing for Control,* pp. 69–70.

18. This point has been emphasized previously by Borrus, Millstein, and Zysman, *U.S.–Japanese Competition,* p. 15: "The shift to the transistor and ultimately to the integrated circuit reshuffled the composition of the leading component manufacturers. Few of the leading producers of the electron tube managed to retain their component market positions in the new technologies. In this reshuffling process, *defense and aerospace procurement created a market incentive for entrepreneurial risk-taking and*

thereby helped to spawn an independent sector of semiconductor component manufacturers" (emphasis added). Their assessment of market share "reshuffling" is based on I. Mackintosh, *Microelectronics in the 1980's* (London: Mackintosh Publications, 1979) p. 66, table II.

19. Borrus, *Competing for Control*, pp. 71–72.

20. D. Webbink, "Staff Report on the Semiconductor Industry" (Washington, D.C.: Federal Trade Commission, Bureau of Economics, 1977), p. 97. Cited in W. Baldwin, *The Impact of Defense Procurement on Competition in Commercial Markets* (Washington, D.C.: Federal Trade Commission, Office of Policy Planning, 1980), pp. 55–56.

21. Borrus, *Competing for Control*, p. 72.

22. Nineteen hundred sixty-two figure from J. Tilton, *International Diffusion of Technology: The Case of Semiconductors* (Washington, D.C.: The Brookings Institution, 1971), cited in *ibid.*; other figures from U.S. Department of Commerce, "Report on the Semiconductor Industry" (1979) (also cited in Borrus, *Competing for Control*, and interview materials).

23. D. Noble, "Social Choice in Machine Design: The Case of Automatically Controlled Machine Tools," in A. Zimbalist, ed., *Case Studies on the Labor Process* (New York: Monthly Review Press, 1979), p. 25. MIT was chosen because of the Servo Lab's experience with automated gunfire control systems.

24. Although military and managerial preferences for top-down control certainly may have played an important role in promoting acceptance and enthusiasm of NC and APT, technological development was driven primarily by the Air Force's specific end-use requirements. Noble acknowledges, in fact, that "Air force performance specifications for four- and five-axis machining of complex parts, often out of difficult materials were simply beyond the capacity of either record-playback (or manual) methods." Noble, "Social Choice in Machine Design," p. 29.

25. Faced with a cutback in Air Force support for the numerical control project in the spring of 1955, Servo Lab engineers, who had developed a program for two-dimensional two-axis machining, pushed for new funding to complete work on a "more efficient" program suited to three-dimensional, three-axis machining. Hoping to reap the benefits of its previous investment, the Air Force obliged, but not before defining an even more forward-looking standard for five-axis control. See D. Noble, *Forces of Production: A Social History of Industrial Automation* (New York: Oxford University Press, 1984), pp. 140–41.

26. *Ibid.*, p. 143.

27. C. Sabel, *Work and Politics: The Division of Labor in Industry* (Cambridge [Cambridgeshire]; New York: Cambridge University Press, 1982), p. 69. Or, in the words of Harley Shaiken, APT, for most metalworking operations, was the equivalent of "using an M-1 tank to drive to work." Shaiken, *Work Transformed: Automation and Labor in the Computer Age* (New York: Holt, Rinehart, & Winston, 1984), p. 100. A 1981 survey indicated that "while 46% of the firms with a large number of NC machine tools (11 or more) used APT, only 15% of the firms with a medium number (5–10) and 13% of the firms with a small number (less than 5) used APT." P. Ong, "NC Machine Tools," in Industry and Trade Strategies, "Programmable Automation Industries," Report to the Congressional Office of Technology Assessment (Contract no. 333-2840), April 1983. Nevertheless, because the initial use of APT created a set of software programs that could not easily be translated, it continued to be the de facto industry standard long after a new generation of simpler programming languages became commercially available.

28. As Noble states, "Companies that wanted military contracts were compelled to adopt the APT system, and those who could not afford the system, with its training requirements, its computer demands, and its headaches, were thus deprived of government jobs." Noble, "Machine Design," p. 28.

29. *Ibid.*, p. 25; A. DiFilippo, *Military Spending and Industrial Decline: A Study of the American Machine Tool Industry* (Westport, CT: Greenwood Press, 1986), p. 57.

30. Over the years, in fact, the Air Force made a practice of favoring with its contracts the core of suppliers it created in the 1950s, many of whom remained major producers into the 1990s. Ong, "NC Machine Tools."

31. For a similar view, see D. Collis, "The Machine Tool Industry and Industrial Policy, 1955–82," in Spence and Hazard, eds., *International Competitiveness* (Cambridge, Mass.: Ballinger, 1988), p. 108.

32. Within many of these companies, engineers had developed in-house languages to program NC equipment; they were typically less flexible than APT, but simpler to use.

33. Noble, "Machine Design," note, pp. 27–28; Noble, *Forces of Production,* p. 209.

34. For more description and analysis of the Japanese machine tool industry, see Ong, "NC Machine Tools"; E. Vogel, *Comeback: Building the Resurgence of American Business* (New York: Simon & Schuster, 1985); M. Fransman, "International Competitiveness, Technical Change, and the State: The Machine Tool Industry in Taiwan and Japan," *World Development* 14, no. 12 (1986); D. Friedman, *The Misunderstood Miracle: Industrial Development and Political Change in Japan* (Ithaca, N.Y.: Cornell University Press, 1988); and Collis, "The Machine Tool Industry."

35. Ong, "NC Machine Tools."

36. Fransman, "International Competitiveness," p. 1383.

37. Ong, "NC Machine Tools."

38. Vogel, *Comback,* p. 80, and Fransman, "International Competitiveness," pp. 1382–83; both cited in D. Gold, "The Impact of Defense Spending on Investment, Productivity, and Economic Growth" (Washington, D.C.: Defense Budget Project, 1990), p. 52.

39. Melman, *Profits without Production,* pp. 8–10.

40. U.S. Department of Commerce, International Trade Administration, "A Competitive Assessment of the U.S. Manufacturing Automation Equipment Industries" (June 1984), pp. 22–25, 36–37.

41. U.S. Department of Commerce, "Metalworking Equipment," *U.S. Industrial Outlook 1985* (Washington, D.C., 1985), Chapter 21, pp. 21–27.

42. C. Chien and L. Conigliaro, "Machine Tools Industry Update" (Prudential-Bache Securities, November 4, 1985).

43. Defense Science Board Task Force, "Report on Industrial Competitiveness" (Richard Furhman, chairman), November 21, 1980; the Air Force Systems Command statement on defense industrial base issues (General Alton Slay), November 21, 1980; House Armed Services Committee Industrial Base Panel Report, "The Ailing Defense Industrial Base: Unready for Crisis" (Richard Ichord, chairman), December 31, 1980. See also the influential book by J. S. Gansler, former deputy assistant secretary of defense and assistant director of defense research and engineering: *The Defense Industry* (Cambridge, Mass.: MIT Press, 1980).

44. Emerging as it did in an increasingly conservative political climate, the DIB concept gave politically tenable expression to an interventionist enthusiasm somewhat sheepishly shared by many supporters of Ronald Reagan. Industrialists, venture capitalists, university presidents, Wall Street bankers, and significant segments of the military establishment all stood to benefit from increased government assistance to the high-

technology industry. Nonetheless, most were tied to an ideological stance that denied the legitimacy of state action in the economy. Accordingly, these diverse economic interests began to coalesce around the political cause of strengthening the "defense industrial base." For more details, see D. Dickson, *The New Politics of Science* (Chicago: University of Chicago Press, 1988).

45. Cited in *ibid*. In retrospect, of course, this report seems to have seriously overestimated Soviet capabilities.

46. Committee on Assessment of the Impact of the DOD Very High Speed Integrated Circuit Program, *An Assessment of the Impact of the Department of Defense Very High Speed Integrated Circuit Program* (Washington, D.C.: National Academy Press, January 1982), p. 6, and interview materials.

47. By the late 1970s, only 5 percent of the industry's R&D expenditures were being funded through military contracts. R. Davis, *IEEE Computer* (July 1979); Committee on Assessment, *An Assessment of the Impact,* p. 22; and Semiconductor Industry Association estimates.

48. Quoted by K. Julian in *High Technology* (May 1985), pp. 49–57.

49. The military services were more interested in high-speed than large-scale processing.

50. Testimony of William J. Perry to the U.S. Congress, Senate Committee on Armed Services, *Hearings on Department of Defense Authorization for Appropriations for Fiscal Year 1980*, 96th Cong., 1st sess.

51. Interview materials. For a similar finding, see G. Fong, "The Potential for Industrial Policy: Lessons from the Very High Speed Integrated Circuit Program," *Journal of Policy Analysis and Management 5*, no. 2 (1986). Many semiconductor firms—most prominent among them, Intel—saw no valuable complementarities or spillover potential in VHSIC and so chose not to participate. In all, ten merchant semiconductor firms eventually sought participation in the VHSIC program.

52. Interview materials and materials supplied by the VHSIC program office.

53. M. Yoshino and G. Fong, "The Very High Speed Integrated Circuit Program: Lesson for Industrial Policy," in B. Scott and G. Lodge, eds., *U.S. Competitiveness in the World Economy* (Boston: Harvard Business School Press, 1986), p. 182.

54. Materials supplied by the VHSIC program office. Also cited in Fong, "The Potential for Industrial Policy."

55. See G. Adams, *The Iron Triangle: The Politics of Defense Contracting* (New Brunswick, N.J.: Transaction Books, 1982), especially pp. 165–73, for more on the many advisory boards that link the broader scientific and industrial communities to the U.S. DOD. My thanks to Glenn Fong for pointing out the importance to VHSIC of the 1978 AGED reviews.

56. The sessions were also attended by representatives of Bell Labs, Cal Tech, Carnegie-Mellon, Clemson, Cornell, Fairchild, Hewlett-Packard, the Institute for Defense Analysis, Jet Propulsion Labs, Johns Hopkins, Lincoln Labs, MIT, RCA, Research Triangle Institute, SRI, Stanford, Tektronix, Texas Instruments, TRW, the University of California at Berkeley, and Westinghouse.

57. *Aviation Week and Space Technology* (February 16, 1981). Cited by L. Brueckner and M. Borrus, "Assessing the Commercial Impact of the Pentagon's Very High Speed Integrated Circuit Program" (BRIE Working Paper No. 5, University of California, Berkeley, November 1984) p. 29.

58. D. Moore and W. Towle, "The Industry Impact of the Very High Speed Integrated Circuit Program: A Preliminary Analysis" (Arlington: Analytic Sciences Corporation, 1980).

59. Interview materials. Fong, "The Potential for Industrial Policy," reports similar findings. This practice was not universal, however. Brueckner and Borrus, "Assessing the Commercial Impact," found, in one instance, a VHSIC contractor that kept VHSIC and VLSI work strictly separate for legal reasons, even though the work itself was substantially identical (p. 73).

60. The process of getting a chip approved for military use still takes so long and requires so much bureaucratic red tape that the chips slated for use in a weapons system often are obsolete by the time the system makes it from design to production. Military screening often takes more than a year and is responsible, by some accounts, for over half the cost of a typical military-qualified chip. The long lag time encourages military system suppliers to resort to source control drawings (SCDs), a long list of specifications for manufacture and testing that enable contractors to avoid the hassle and expense of getting commercial or dual-use devices approved for military use. Ironically, this has led to a situation in which a specification system designed to standardize the industry has in fact encouraged the proliferation of costly nonstandard chips. The explosion of SCDs impedes quality control efforts as chip producers are overwhelmed with "thousands of separate specifications of devices that are in many cases identical. [The system] produces extremely expensive products that at best are only equal to their commercial, off-the-shelf counterparts and in some cases are *worse*." See A. C. Revkin, "A War over Military Chips," *Science Digest* (July 1985), pp. 56–79.

61. In any event, military project managers often waived military-oriented performance requirements, such as man-rated radiation hardness, when such parameters seemed irrelevant to the performance of the chip (interview materials).

62. Interview materials.

63. Theoretically, we would want to explore the issue of opportunity costs—that is, what would have happened had the resources expended on VHSIC been expended, instead, in their best alternative commercial use. In the real world, however, there is no evidence to indicate that a denial of funds to VHSIC would have resulted in comparable government (or private) spending on the development of civilian applications.

64. TI was not among those teams awarded one of three Phase II contracts in October 1984, for reasons that remain unclear, since the Pentagon did not publicly disclose its criteria for selection. The Phase II "winners" were IBM, TRW, and Honeywell. Julian, *High Technology*, p. 53.

65. Efficient use of the chip's "real estate" may be less important for custom applications, where the number of custom designs available from a firm may be the crucial competitive variable. But, in that case, VHSIC can have positive commercial effects only if computer-aided design tools developed by the military are characterized by open architectures that can quickly and flexibly be adapted to civilian uses. Quick turnaround technology is certainly important, but even dramatic improvements in military turnaround times (say, from seven down to two years) are nowhere near the speeds required by custom chip producers in the commercial marketplace. See Brueckner and Borrus, "Assessing the Commercial Impact," pp. 47–48.

66. Like Texas Instruments, Phase I contractor IBM also served both military and civilian markets, but unlike TI, IBM did not sell semiconductors on the open market. IBM pursues its civilian and military lines of business in strictly separate facilities.

67. Interview materials.

68. Interview materials.

69. Betac Corporation, "Final Report: Phase 2 of the United States Technology Transfer Export Controls Project" (Arlington: Betac, January 1980). Quoted in U.S.

Congress, House Committee on Government Operations, *The Classification of Private Ideas*, Report No. 96–1540 (December 22, 1980).

70. Julian, *High Technology*, p. 57.

71. Another reason that firms began to compartmentalize their military-sponsored R&D was the application of *criminal* penalties to the commercial use of Pentagon money. The criminalization of such activity was Secretary of Defense Weinberger's response to congressional furor over a number of highly publicized cost overruns—for example, the Pentagon's $700 coffee pot—that seemed, for a time, to threaten the Administration's military build-up. Interview materials.

72. Brueckner and Borrus, "Assessing the Military Impact," p. 73.

73. Quoted in Julian, *High Technology*, p. 57.

74. *Electronic Business* (February 6, 1989), p. 56.

75. E. D. Maynard, Jr., director of VHSIC and electron devices in the Office of the Undersecretary of Defense for Research and Engineering. Quoted by B. Karlin, *Electronic Business* (August 1, 1986), p. 72.

76. "Expert systems" were themselves spun off from rule-base programming work sponsored by DARPA in the late 1970s. So were UNIX-based workstations, which emerged in 1982 from a DARPA-sponsored project at Stanford University, and "Berkeley 4.2," developed at the University of California at Berkeley. This apparently successful spin-off again exhibits traits of what we would now want to characterize as spin-on—the original UNIX software was developed for *commercial* purposes; subsequently, ARPANET served as a bridge for further co-development by military and commercial users whose needs for developing network communications among different computer systems increasingly converged.

77. Japan's Fifth Generation project was designed to develop commercial applications of the same technologies targeted for development by SCP, over approximately the same time frame. The Japanese were expected to spend approximately $500 million on the ten-year Fifth Generation project, plus another $200 million on a separate five-year Superspeed Computer project. Japan's government-backed efforts in artificial intelligence and supercomputing are focused explicitly on the enhancement of business and consumer productivity and on the improvement of social services. *Datamation* (August 1, 1984), p. 42.

78. The decision greatly disturbed many computer researchers who appreciated the financial support, but who felt, nevertheless, that SCP's emphasis on showcase projects for each of the armed forces would take too much money and effort away from essential basic research. For more details on the computer science community's initial reaction to SCP, see J. Stowsky, "Beating Our Plowshares into Double-Edged Swords: The Impact of Pentagon Policies on the Commercialization of Advanced Technologies," BRIE Working Paper No. 17 (April 1986), pp. 54–59.

79. See W. Schatz and J. W. Verity, "Weighing DARPA's AI Plans," *Datamation* (August 1, 1984), pp. 34–43; and Schatz and Verity, "DARPA's Big Push in AI," *Datamation* (February 1984), pp. 48–50.

80. Interview materials. See also *Datamation* articles previously cited.

81. Interview materials. See also D. B. Davis, "Super Computers: A Strategic Imperative?" *High Technology* (May 1984), pp. 44–52; U.S. Congress, Office of Technology Assessment, *Information Technology R&D: Critical Trends and Issues* (Washington, D.C.: U.S. Government Printing Office, February 1985), pp. 57–62; R. Corrigan, "The Latest Target of the Japanese—U.S. Preeminence in Supercomputers," *National Journal* (April 2, 1983), pp. 688–92; N. R. Miller, "Supercomputers," *CRS Review* (Congres-

sional Research Service) (March 1984), pp. 17–19; and see "Supercomputing: Number-Crunching for Research," in *Physics Today* (May 1985), pp. 51–53. Others in the computer field were, of course, quite sanguine about the prospects for substantial overlap between advances in knowledge-based software and superspeed computing. "There are some people who are jealous of the large amount of money DARPA is spending on artificial intelligence supercomputing to the exclusion of scientific supercomputing," said Burton J. Smith, vice president in charge of R&D at Denelcor, one of the three U.S. supercomputer manufacturers. But, he added, "I think the artificial intelligence work will prove very beneficial to superspeed computing and vice versa." Quoted in Davis, "Super Computers," p. 47.

82. *Datamation* articles previously cited. Plus A. Pollack, "Pentagon Sought Smart Truck But It Found Something Else," *New York Times* (May 30, 1989), p. 1.

83. Pollack, *ibid.*, p. 1, and interview materials.

84. Michael R. Leibowitz, "Does Military R&D Stimulate Commerce or Pork Barrel?" *Electronic Business* (February 6, 1989), pp. 54–58.

85. By 1987, the vehicle was supposed to be able to travel across six miles of open desert at speeds up to three miles an hour, avoiding bushes and ditches along the way. In fact, it was only able to travel about 600 yards at about two miles an hour, slower than many people walk. Pollack, "Pentagon Sought Smart Truck," p. 1.

86. Theodore H. Moran, "The Globalization of America's Defense Industries: Managing the Threat of Foreign Dependence," *International Security,* Vol. 15, no. 1, Summer 1990.

Chapter 5

1. I am indebted to Etel Solingen, Gene Rochlin, John Holdren, and John Zysman for suggestions and comments on an earlier version of this chapter.

2. As S. J. Lundin points out, such allegations are controversial and in large measure unsubstantiated. Other than occasional references to Iran, Iraq, Syria, and Libya, U.S. assertions have not been accompanied by the names of specific nations thought to have, or to be developing, such capabilities. See "Chemical and Biological Warfare: Developments in 1989," in Stockholm International Peace Research Institute (SIPRI), *World Armaments and Disarmament: SIPRI Yearbook 1990* (London: Oxford University Press, 1990).

3. Among these countries, only India has reached the point of successfully launching a satellite. On ballistic missile programs in the Third World, see Aaron Karp, "Ballistic Missile Proliferation," in SIPRI, *World Armaments and Disarmament*; Janne Nolan, "Ballistic Missiles in the Third World: The Limits of Non-Proliferation," *Arms Control Today* 19 (November 1989), pp. 9–14.

4. For a summary of the regime's provisions, see F. Hollinger, "The Missile Technology Control Regime: A Major New Arms Control Agreement," in U.S. Arms Control and Disarmament Agency (ACDA), *World Military Expenditures and Arms Transfers 1987* (Washington: ACDA, 1988).

5. M. Brzoska and T. Ohlson, eds., *Arms Production in the Third World* (London: Taylor and Francis, 1986).

6. Calculated from data in SIPRI, *World Armaments and Disarmament*, pp. 252–53 (Table 7A.2). Exports by Third World producers were calculated by adding the SIPRI figures for China, Israel, Brazil, and "Other Third World."

7. See Brzoska and Ohlson, eds., *Arms Production,* Table 2.3, p. 16.

8. Calculated from *ibid.*, appendix 1.

9. Estimated defense spending in less developed countries was $173 billion in 1985, according to the U.S. ACDA. Although this represented a slightly smaller percentage of GNP dedicated to defense than in the case of the industrial countries, it also represented a 25 percent constant-dollar increase during the preceding decade. Ruth Leger Sivard estimates that 1988 Third World expenditures for defense ($134 billion) were almost equal to those for health and education combined ($147 billion). See U.S. ACDA, *World Military Expenditures*; Sivard, *World Military and Social Expenditures 1991* (Washington, D.C.: World Priorities Inc., 1991).

10. Chung-in Moon, "South Korea: Between Security and Vulnerability," in J. Katz, ed., *The Implications of Third World Military Industrialization: Sowing the Serpent's Teeth* (Lexington, Mass.: D.C. Heath, 1986). See also Janne Nolan, *Military Industry in Taiwan and South Korea* (New York: St. Martin's Press, 1986).

11. "McDonnell to Supply Jets to Seoul," *New York Times*, December 21, 1989, p. C-1.

12. On Israel, see A. Klieman, "Middle-Range Arms Suppliers: The Israeli Case," *Journal of International Affairs* 40 (Summer 1986), pp. 115–28; G. Steinberg, "Indigenous Arms Industries and Dependence: The Case of Israel," *Defense Analysis* 2 (December 1986), pp. 291–305. On Taiwan, see Nolan, *Military Industry*; D. Louscher and M. Salamone, *Technology Transfer and U.S. Security Assistance: The Impact of Licensed Production* (Boulder, Colo.: Westview Press, 1986).

13. See V. Millan, "Argentina: Schemes for Glory," in Brzoska and Ohlson, eds., *Arms Production*; M. Canoura, "A Importância da Indústria Bélica Para a Segurança Nacional," *Política e Estratégia* 6 (1988), pp. 363–73. See also "Argentina Takes Steps to Boost Growth of Aerospace Industry," *Aviation Week and Space Technology*, August 17, 1987; "New Military Industrial Complex Proposed," *Buenos Aires Herald*, March 21, 1986.

14. Raju Thomas, *Indian Security Policy* (Princeton, N.J.: Princeton University Press, 1986), p. 218.

15. *Ibid.*, pp. 218–19. The term *structural unity* is quoted by Thomas from K. Subrahmanyam, "Planning Defence Production: Integral to Industrial Growth," *Times of India*, January 24, 1980.

16. W. Frieman, "China's Military R&D System: Reform and Reorientation," in D. Simon and M. Goldman, *Science and Technology in Post-Mao China* (Cambridge, Mass.: Harvard University Press, 1989).

17. Michael Renner, "Swords into Plowshares: Converting to a Peace Economy," Worldwatch Paper 96, Worldwatch Institute, Washington, D.C., June 1990.

18. M. Dunn, "Egypt: From Domestic Needs to Export Market," in Katz, ed., *The Implications*. See also Louscher and Salamone, *Technology Transfer*.

19. D. Weatherbee, "Indonesia: Its Defense-Industrial Complex," in Katz, ed., *The Implications*.

20. A. Karp, "Controlling the Spread of Ballistic Missiles to the Third World," *Arms Control* 7 (May 1986), pp. 31–46. See also R. Shuey, "Missile Proliferation: Survey of Emerging Missile Forces," Congressional Research Service, Washington, D.C., October 3, 1988; Nolan, "Ballistic Missiles in the Third World."

21. See M. Castells, "High Technology, Economic Policies and World Development," Working Paper No. 18, Berkeley Roundtable on the International Economy, University of California–Berkeley, May 1986.

22. The initial challenge to traditional suppliers was particularly strong in Latin

America. See Michael Klare, *American Arms Supermarket* (Austin: University of Texas Press, 1984), Chap. 5.

23. Ten countries accounted for 72 of 100 arms technology licensing transactions recorded by The SIPRI for 1977–83. See D. Louscher and M. Salamone, "Brazil and South Korea: Two Cases of Security Assistance and Indigenous Production Development," in Louscher and Salamone, eds., *Marketing Security Assistance: New Perspectives on Arms Sales* (Lexington, Mass.: D. C. Heath, 1987).

24. Stephen Mark Haggard, *Pathways from the Periphery: The Politics of Growth in the Newly Industrializing Countries* (Ithaca, N.Y.: Cornell University Press, 1990), p. 254.

25. The most complete analyses of the Brazilian defense sector have been done by Brazilians. See C. Brigagão, *O Mercado da Segurança* (Rio de Janeiro: Editora Nova Fronteira, 1984); R. Dagnino, "A Indústria de Armamentos Brasileira: Uma Tentativa de Avaliaçao" (Ph.D. diss., Economics Institute, State University of Campinas, Brazil, August 1989); J. Drumond Saraiva, "O Desenvolvimento Industrial Bélico" (monograph, July 1989). In English, see C. Brigagão, "The Brazilian Arms Industry," *Journal of International Affairs* 40 (Summer 1986), pp. 101–14; A. Barros, "Brazil," in Katz, ed., *Arms Production in Developing Countries: An Analysis of Decision-Making* (Lexington, Mass.: D. C. Heath, 1984); P. Lock, "Brazil: Arms for Export," in Brzoska and Ohlson, eds., *Arms Production*; P. Franko-Jones, "Public-Private Partnership: Lessons from the Brazilian Armaments Industry," *Journal of Interamerican Studies and World Affairs* (Winter 1987–88), pp. 41–66; Louscher and Salamone, "Brazil and South Korea"; W. Perry and J. C. Weiss, "Brazil," in J. Katz, ed., *The Implications*; E. Kapstein, "The Brazilian Defense Industry and the International System," *Political Science Quarterly* 105 (Winter 1990–91), pp. 579–96.

26. "Brazil Forms High-Technology Venture to Develop Advanced Weapons Systems," *Aviation Week and Space Technology*, August 17, 1987; "Embraer se associa à outras empresas para produzir armas," *O Estado de São Paulo*, December 23, 1986.

27. Development of a launch vehicle, which would also yield a crude ballistic missile capability, appears to have been slowed by the technology-transfer restrictions of the Missile Technology Control Regime (MTCR) adopted by seven supplier governments in 1987. Reports of French willingness to transfer sensitive technology, however, and Brazil's 1988 agreement with nonregime member China, raise questions about the regime's effectiveness. On the MTCR, see Hollinger, "The Missile Technology Control Regime"; on Brazilian-French cooperation, see "U.S. Seeks to Stop Brazil Deal to Gain Missile Technology," *New York Times*, October 19, 1989; "França garante a cessao ao Brasil de tecnologia espacial," *O Globo*, October 6, 1989.

28. "Brazil, China Form Space Launch Venture," *Aviation Week and Space Technology*, May 25, 1989.

29. D. Albright, "Bomb Potential for South America," *Bulletin of American Scientists*, May 1989. See also the report of the Brazilian Physics Society (Sociedade Brasileira de Física), "A viabilidade de enriquecer urânio no Brasil em grau para bomba e o projeto conceitual de um explosivo nuclear" (Rio de Janeiro: Sociedade Brasileira de Física, May 1990).

30. On the German–Brazilian nuclear accord, see N. Gall, "Atoms for Brazil, Dangers for All," *Foreign Policy* 23 (Summer 1976), pp. 155–201; D. Myers, "Brazil: Reluctant Pursuit of the Nuclear Option," *Orbis* 27 (Winter 1984), pp. 881–911.

31. These figures are from Brigagão, "The Brazilian Arms Industry."

32. The employment estimate is from Saraiva, "O Desenvolvimento Industrial

Bélico." Dagnino "A Indústria," presents an upper-bound estimate of total (direct plus indirect) employment of 100,000 workers.

33. Calculated from data in the World Bank's *World Development Report 1988* (Washington, D.C.: World Bank, 1988). Of total 1986 merchandise exports of $22.4 billion, 41 percent ($9.2 billion) were manufacturing exports, of which 43 percent ($3.9 billion) were destined for less developed countries, including oil-exporting nations.

34. The estimate is that of Dagnino, "A Indústria," which includes the total of arms exports by the three leading defense-sector firms (Avibrás, Engesa, and Embraer; Embraer data excludes sales of civilian aircraft). Dagnino argues that these three firms account for the vast majority of Brazilian arms sales. Dagnino's estimates are roughly consistent with those of SIPRI, which as published include only exports of so-called major weapons systems to Third World buyers. SIPRI data suggest that Brazilian exports of major weapons systems to Third World buyers peaked at $491 million (in 1985 dollars) in 1987. See SIPRI, *World Armaments and Disarmament*, p. 253.

35. On the origins of the Brazilian defense sector, see F. McCann, Jr., "The Brazilian Army and the Pursuit of Arms Independence, 1899–1979," in B. F. Cooling, ed., *War, Business and World Military Industrial Complexes* (Port Washington, N.Y.: Kennikat Press, 1981); S. Hilton, "The Armed Forces and Industrialists in Modern Brazil: The Drive for Military Autonomy," *Hispanic American Historical Review* 62 (November 1982), pp. 629–73; Brigagão, *O Mercado*; Dagnino, "A Indústria."

36. On the political and economic role of the military during this period, see John D. Wirth, *The Politics of Brazilian Development, 1930–1954* (Stanford, Calif.: Stanford University Press, 1970); José Murilo de Carvalho, "Armed Forces and Politics in Brazil, 1930–45," *Hispanic American Historical Review* 62 (May 1982), pp. 193–223. For a dissenting view, see Stanley Hilton, "Military Influence on Brazilian Economic Policy, 1930–1945: A Different View," *Hispanic American Historical Review* 53 (February 1973), pp. 71–94.

37. See General Pedro Aurelio Góes Monteiro, *A Revolução de 30 e a Finalidade Política do Exército* (Rio de Janeiro: Adersen Editores, n.d.), p. 133, cited in Edmundo Campos Coelho, *Em Busca de Identidade: O Exército e a Política na Sociedade Brasileira* (Rio de Janeiro: Forense-Universitária, 1976), p. 103 (my translation).

38. For varying points of view on the role of the ESG, see A. Stepan, *The Military in Politics: Changing Patterns in Brazil* (Princeton, N.J.: Princeton University Press, 1971); Campos Coelho, *Em Busca de Identitade*; J. Markoff and S. Baretta, "Professional Ideology and Military Activism in Brazil: Critique of a Thesis of Alfred Stepan," *Comparative Politics* 17 (January 1985), pp. 175–91; F. McCann, Jr., "Origins of the 'New Professionalism' of the Brazilian Military," *Journal of Interamerican Studies and World Affairs* 21 (November 1979), pp. 505–22.

39. For a discussion of civil-military relations in the context of the military regime's economic policy, see L. Bresser Pereira, *Development and Crisis in Brazil, 1930–1983* (Boulder, Colo.: Westview, 1984).

40. The late 1950s saw a wave of "democratization" on the South American continent similar to that of the 1980s, with only the military-authoritarian regime of Paraguay's General Stroessner remaining by the early 1960s. This trend was reversed by military coups in Brazil (1964), Argentina (1966), Ecuador (1966), Peru (1968), Chile (1973), Uruguay (1973), and a series of coups during this period in Bolivia.

41. Brigagão, "The Brazilian Arms Industry."

42. On state support for the defense sector, see R. Dagnino, "Indústria de armamentos: o Estado e a tecnologia," *Revista Brasileira de Tecnologia* 14 (May/June 1983),

pp. 5–17; Franko-Jones (1987–88); R. Hudson, "The Brazilian Way to Technological Independence: Foreign Joint Ventures and the Aircraft Industry," *Inter-American Economic Affairs* 37 (1983), pp. 23–44. On the role of state autonomy in policy coordination and implementation, see Etel Solingen, "State Autonomy, Lateral Autonomy, and Sectoral Adjustments: Arms and Nuclear Industries in Brazil and Argentina." (Paper presented at the International Studies Association Meeting, London 1989.)

43. On the formation of Embraer, see Dagnino, "Indústria de armamentos"; Hudson, "The Brazilian Way"; Ozires Silva, "O vôo da Embraer," *Revista Brasileira de Tecnologia* 13 (January–March 1982), pp. 20–30.

44. When Embraer celebrated its twentieth anniversary in 1990, its management team consisted of several former CTA personnel. For example, graduates of ITA, the Air Force's engineering school located within the CTA complex at São José dos Campos, include the chairman of the Embraer board, the president, 5 of 6 directors, over half of the firm's 50 managers, and 40 percent of its roughly 1,200 engineers. Source: interviews with Embraer officials, São José dos Campos and Rio de Janeiro, Brazil, August–September 1990.

45. Hudson, "The Brazilian Way." See also "Brazil's Air Force Purchases Linked to Embraer Pacts," *Aviation Week and Space Technology*, June 25, 1984.

46. A. Schwarz, "Arms Transfers and the Development of Second-Level Arms Industries," in Louscher and Salamone, eds., *Marketing Security Assistance*. This estimate excludes indirect technological gains acquired by importing completed weapons systems.

47. Franko-Jones, "Public-Private Partnership." See also Maria Carlotta de Souza Paula, "Aeronaves: Os rumos da indústria brasileira," *Revista Brasileira de Tecnologia* 16 (May–June 1985), pp. 48–56; Dagnino, "Indústria de armamentos."

48. Bresser Pereira, *Development and Crisis*, particularly Chapters 8 and 9; Solingen, *State Autonomy*.

49. The term public-private partnership has been used by Franko-Jones, "Public-Private Partnership," in her analysis of the firms Engesa, Embraer, and Avibrás.

50. For an elaboration of this point, see Dagnino, "A Indústria," Chapter 6.

51. "Dívida da indústria bélica supera US$ 1 bi," *Folha de São Paulo*, January 21, 1990.

52. On Avibrás, see "Sales Halved," *Gazeta Mercantil* (English-language edition), January 30, 1989; "Filing For Concordata," *Gazeta Mercantil* (English-language edition), January 15, 1990; "Concordata da Avibrás foi surpresa," *Folha de São Paulo*, January 21, 1990. On Engesa, see "Tank Fatigue," *Gazeta Mercantil* (English-language edition), August 21, 1989.

53. "Huge Sums Needed for New Aircraft," *Gazeta Mercantil* (English-language edition), August 28, 1989; "Turbulência no ar," *Veja*, June 14, 1989.

54. "Brazilian Plane Maker Embraer, Once a Symbol of Third-World Strength, Puts Its Hopes on Hold," *Wall Street Journal*, November 13, 1990.

55. After more than two decades of military rule, a civilian president was chosen by the Brazilian Congress in 1985. A new constitution, based largely on principles of representative democracy, was put into effect in October 1988, with the first direct presidential election in twenty-nine years held in December 1989. On the transition from authoritarian rule, see G. O'Donnell and P. Schmitter, *Transitions from Authoritarian Rule: Tentative Conclusions about Uncertain Democracies* (Baltimore: Johns Hopkins University Press, 1986); W. Selcher, ed., *Political Liberalization in Brazil: Dynamics, Dilemmas, and Future Prospects* (Boulder, Colo.: Westview, 1986); A. Stepan, *Rethink-*

ing Military Politics: Brazil and the Southern Cone (Princeton, N.J.: Princeton University Press, 1988); G. O'Donnell, "Challenges to Democratization in Brazil," *World Policy Journal* 5 (Spring 1988), pp. 281–300.

56. Some observers have cited the poor performance of the Argentine military during the 1982 Malvinas (Falklands) conflict as the onset of this renewed preoccupation with external defense. The Malvinas war generated deep concerns within the Brazilian armed forces as to their own state of readiness; several analyses of the implications of the conflict appeared in Brazilian military journals. See S. Hilton, "The Brazilian Military: Changing Strategic Perceptions and the Question of Mission," *Armed Forces and Society* 13 (Spring 1987), pp. 329–51. The fact that most of the emphasis on redefining the military's role has come from the Navy and Air Force, which would inevitably gain in budget and influence at the Army's expense in any such shift, suggests the important role of institutional political considerations, however.

57. For a discussion of the role of the defense sector in the political transition and the impact of the transition on defense-sector politics, see Ken Conca, "Technology, the Military, and Democracy in Brazil," forthcoming in *Journal of Interamerican Studies and World Affairs*.

58. On the changing structure of the global arms economy in the 1980s, see M. Klare, "The State of the Trade: Global Arms Transfer Patterns in the 1980s," *Journal of International Affairs* 40 (Summer 1986), pp. 1–21; M. Klare, "Deadly Convergence: The Perils of the Arms Trade," *World Policy Journal* 6 (Winter 1989), pp. 141–68; A. Kolodziej, *Making and Marketing Arms* (Princeton, N.J.: Princeton University Press, 1987); Louscher and Salamone, eds., *Marketing Security Assistance*.

59. "Na barreira do som: Embraer sonha com o seu caça supersônico," *Isto É*, November 5, 1988. See also "South American Companies Seek Formula to Assure Sustained Growth," *Aviation Week and Space Technology*, August 17, 1987; "Embraer Moves to Solidify Role as Leading Aircraft Manufacturer," *Aviation Week and Space Technology*, August 17, 1987.

60. "Engesa propõe alugar tanques Osório ao Exército," *O Globo*, November 12, 1985.

61. "Filing for concordata," *supra*.

62. "Future uncertain for AM-X Project," *Gazeta Mercantil* (English-language edition), October 23, 1989; "Budget Cut Curtails AM-X," *Foreign Broadcast Information Service* (Latin America), December 12, 1989.

63. Alfred Stepan has argued that the prospects for the Brazilian military to claim budget resources may, paradoxically, be enhanced under democratic civilian rule. See Stepan, *Rethinking Military Politics,* Chap. 6.

64. The production agreement for the AM-X tactical jet fighter, involving Embraer and the Italian firms Aermacchi and Aeritalia, illustrates the endurance of this intergenerational dependence. Embraer is responsible for 30 percent of the work and development costs; its contributions are limited to the wings, empennage, pylons,and fuel tanks. The Italians are responsible for the front, center,and rear fuselage, as well as the carbon fiber components of the wing and empennage. The engine is produced under license from Rolls Royce. See "AM-X Fighter Prototype Makes First Flight," *Aviation Week and Space Technology*, May 21, 1984; "Embraer AM-X Prototype Makes First Flight in Brazil in Preparation for Test Program," *Aviation Week and Space Technology*, November 4, 1985.

65. Offshoring is unlikely to be driven by comparative advantage, given Brazil's combination of a developed industrial infrastructure and low wage rate for skilled labor. Off-

shoring could be driven, however, by the increasing leverage buyers exert in a weak market, and the likely demands for technology transfer associated with that leverage.

66. Dagnino, "A Indústria," p. 347.

67. The latest generation of Brazilian submarines, designed by the German firm IKL and under construction at the Navy's Rio de Janeiro shipyard, have an index of nationalization of less than 30 percent; in terms of installed equipment, only the batteries are produced in Brazil. And the Inhauma-class corvettes currently under construction, although 100 percent Brazilian in terms of design and assembly, have an index of nationalization of 47 percent, with imported radar, sonar, missiles, cannon, and other arms-system components. In both cases, Navy officials see little chance for increasing nationalization of production without significant increases in scale. See *Revista Brasileira de Tecnologia*, October 1989.

68. A classified 1987 intelligence report of the West German government, later leaked to the German press, concluded that such diversions had in fact occurred (David Albright, Federation of American Scientists, personal communication, October 1989).

69. Brazil unilaterally canceled its long-standing defense-cooperation accords with the United States in 1977, in response to U.S. pressure on human rights and nuclear proliferation concerns. An accord on military-technological cooperation was signed in 1984, but its implementation has been frustrated by Brazil's unwillingness to accept U.S. restrictions on retransfer of technologies and products. See "U.S.–Brazil Arms Pact Opposed," *Financial Times*, February 21, 1984.

70. "Government to Continue Supplying Arms to Libya" (Madrid: EFE, 1647 GMT, January 9, 1986) (cited in FBIS Latin America, January 10, 1986); "Heavy Weapons Will Be Sold to Libya" (Paris: AFP, 1810 GMT, August 4, 1986) (cited in FBIS Latin America, August 5, 1986).

71. "Cientista das Arábias," *Veja*, October 3, 1990.

72. A high-ranking Air Force official, for example, suggested that the service's two priorities of the mid-1980s, the AM-X project and expansion of the nation's air traffic control system, "were just too much." See "Brazil's Goal of Self-Sufficiency in Arms Impeded by Inflation, Record High Indebtedness," *Aviation Week and Space Technology*, August 24, 1987, p. 40.

73. A. Britto de Castro, A. Reubens, N. Majlis, L. Pinguelli Rosa, and F. Barros, "Brazil's Nuclear Shakeup: Military Still in Control," *Bulletin of the Atomic Scientists* 45 (May 1989), pp. 22–25.

74. See "O espaço deve ser civil," *Veja*, February 8, 1989 (an interview with fired INPE director Marco Antonio Raupp).

75. "U.S. Debates Selling Supercomputers to 3 Nations," *New York Times*, August 20, 1989. The United States later agreed, during President Bush's December 1990 visit to Brazil, to allow the sale of an IBM supercomputer to Embraer (perhaps in response to the announcement that month that Brazil and Argentina would develop a system of bilateral safeguards and inspections on their nuclear facilities). As of April 1991, the sale to Embraer remained blocked, however, because the Brazilian government had not provided the required guarantees that use would be restricted to civilian applications. See "A Standoff with Brazil on Computer," *New York Times*, April 12, 1991.

76. The 1979 reorganization of the informatics sector, for example, strengthened the involvement of the National Security Council and the intelligence community in sector policy. (See D. Proenca, Junior, "Tecnologia Militar e Os Militares na Tecnologia: O Caso da Politica Nacional de Informática," Master's thesis, COPPE, Universidade Federal do Rio de Janeiro, November 1987. Similarly, the Institute for Energy and Nuclear

Research (IPEN), where much of the military's nuclear research has occurred, was transferred from state to federal authority in 1982, in anticipation of an opposition victory in the São Paulo gubernatorial campaign later that year. (See Centro Ecumênico de Documentação e Informação, *De Angra a Aramar: Os Militares a Caminho da Bomba* (Rio de Janeiro: CEDI, 1988)). In addition, the Armed Forces General Staff retained control of Brazilian space policy when the National Security Council was disbanded. It has been widely reported that, prior to the transition to civilian rule, negotiations between the opposition Democratic Alliance and the military regime included an explicit agreement that the military's technological programs and industrial activities would remain adequately funded and under military control. See E. Rizzo de Oliveira, "Constituente, Forças Armadas, e Autonomia Militar," in E. Rizzo de Oliveira, G. L. Cavagnari Filho, J. Quartim de Morães, and R. A. Dreifuss, eds., *As Forças Armadas no Brasil* (Rio: Espaço e Tempo, 1987), pp. 174–75.

77. On the efforts of the Brazilian military to lobby the Constituent Assembly during the drafting of a new constitution, see Rizzo de Oliveira, "Constituente"; J. Murilo de Carvalho, "Militares e Civis: Um Debate para além da Constituente," in A. Camargo and E. Diniz, eds., *Continuidade e Mudança no Brasil da Nova República* (São Paulo: Vertice, 1989); R. Dreifuss, *O Jogo da Direita* (Petrópolis: Editora Vozes, 1989).

78. The United States represents by far Embraer's largest market for civilian aircraft. The United States threatened sanctions in response to alleged unfair Brazilian trade practices, hurting Embraer's U.S. sales appreciably in 1988.

79. "Huge Sums Needed for New Aircraft," *supra.*

80. "O espaço deve ser civil," *supra.* See also "Construction Slows Down," *Gazeta Mercantil* (English-language edition), April 3, 1989.

81. It is for this reason that I have avoided referring to a Brazilian military-industrial "complex." I am indebted to Renato Dagnino for this observation.

82. The metalworkers' union of São José dos Campos has been the strongest advocate of conversion. Union head José Luis Gonçalves was quoted in September 1989 as saying, "We demand diversification as the primary way out of the social crisis which the arms industry's crisis has generated." See "In the Dumps," *Gazeta Mercantil* (English-language edition), September 4, 1989. See also various editions of the union publication *Apoio Sindical.* On the U.S. example, see "Lobbying Steps Up on Military Buying as Budget Shrinks," *New York Times*, April 9, 1990.

83. See, for example, the proceedings of the conference "Estratégia para o Brasil do Seculo XXI," hosted by the Nucleo de Estudos Estratégicos, Universidade Estadual de Campinas, September 20–21, 1989.

84. This was a recurrent theme in recent interviews with naval officers involved in these programs.

85. "40% of Embraer for sale," *Gazeta Mercantil* (English-language edition), June 24, 1991.

86. IMBEL, the Army-owned munitions complex, was designated to hold a minority share in a restructured Engesa. See "Em segredo," *Veja*, March 6, 1991; "Engesa Sale Negotiations Continue," FBIS Latin America, July 8, 1991.

87. As of this writing, it remains unclear whether this trend will be reversed by the recently announced agreement on a bilateral Brazilian–Argentine system of safeguards and nuclear facility inspections ("Argentina and Brazil Renounce Atomic Weapons," *New York Times*, November 29, 1990). Past announcements of nuclear policy reorganization have had the effect of increasing military control, even when couched in terms of greater opening and societal control. (On this point, see Britto de Castro et al., "Brazil's Nuclear Shakeup.") Such could be the case with the acceptance of nonproliferation safe-

guards if, for example, the *quid pro quo* for their acceptance were an expansion of the military's nuclear programs. An interministerial commission, formed by President Collor shortly after his election and tasked with reevaluating nuclear policy, called for just such an expansion. A pronouncement adopting the commission's proposal was delayed indefinitely by the international furor over Brazilian military-technological links to Iraq; see "Collor adia decisão sobre plano nuclear," *Folha de São Paulo*, September 9, 1990.

88. Much of the following discussion draws on Thomas, *Indian Security Policy*.

89. This point is discussed in *ibid.*, pp. 119–34.

90. *Ibid.*, p. 201.

91. *Ibid.*

92. For a discussion of the role such personalistic connections have played in the militarization of Indian science and technology policy, see Dhirendra Sharma, "India's Lopsided Science," *Bulletin of the Atomic Scientists*, May 1991.

93. The following description of the defense sector draws on Moon, "South Korea," and Nolan, *Military Industry*.

94. Government incentives included concessional financing, advance payments, tax credits, tariff exemptions, and other measures. See Moon, "South Korea," p. 249. On the role of U.S. technological assistance, see Louscher and Salamone, "Brazil and South Korea."

95. Cited in Moon, "South Korea," p. 244.

96. For a summary representation of South Korean elite perspectives on these issues, see W. Taylor, ed., *The Future of South Korean–U.S. Security Relations* (Boulder, Colo.: Westview Press, 1989).

97. Moon, "South Korea," p. 259. The seven conglomerates are Samsung, Hyundai, Lucky–Gold Star, Daewoo, Sangyong, Korea Explosive, and Hanjin. See Moon, pp. 256–57.

98. Charles Wolf, Jr., Donald P. Henry, K. C. Yeh, James H. Hayes, John Schank, and Richard L. Sneider, "The Changing Balance: South and North Korean Capabilities for Long-Term Military Competition," Rand Corporation Report R-3305/1-NA, December 1985, p. 40.

99. Moon, "South Korea," p. 250. The author goes on to blame this highly centralized decision-making structure for the poor coordination that has at times characterized defense-sector policies, and for the sector's enduring inefficiencies: "While the concentration of decision-making power helped rapid implementation of defense industrialization by insulating the defense-industrial sector from competing bureaucratic and political claims in the early stage, its intermediate consequences appear to be devastating."

100. A. Schwarz calculates an index of U.S. technological dependence for Korean arms production of 94 percent. See Schwarz, "Arms Transfers."

101. See Patrick O'Connell, "R.O.K.–U.S. Defense Industrial Cooperation," and Dong Joon Hwang, "R.O.K.–U.S. Defense Industrial Cooperation: A New Step in Security Enhancement," both in Taylor, ed., *The Future of South Korean–U.S. Security Relations*.

Chapter 6

1. See Kenneth N. Waltz, *The Theory of International Politics* (Reading, Mass.: Addison-Wesley, 1979).

2. This statement comes with all the usual caveats regarding American power and the sometimes-different preferences of our allies.

3. See Robert Pollard, *Economic Security and the Origins of the Cold War* (New York: Columbia University Press, 1985).

4. This section draws from Steve Weber, *Multilateralism in NATO: Shaping the Postwar Balance of Power, 1945–61* (Berkeley: University of California Press, Institute of International Studies, 1991).

5. That is, both presidents found that the United States was more constrained by Soviet power than they originally believed and that Washington's freedom of action, although substantial, was not unencumbered.

6. This is apparent in both the East European military situation and the economic disaster following the collapse of bilateral CMEA trade between each East European country and the Soviet Union.

7. For a different view, see Mancur Olson and Richard Zeckhauser, "An Economic Theory of Alliances," *Review of Economics and Statistics*, Vol. 48, no. 3, August 1966, pp. 266–79.

8. See John Ruggie, "International Regimes, Transactions, and Change: Embedded Liberalism in the Postwar Economic Order," in Stephen Krasner, ed., *International Regimes* (Ithaca, N.Y.: Cornell University Press, 1983).

9. Steve Weber accounts for these ideas as part of a desire for transforming a bipolar struggle into multipolarity, reflecting a traditional American faith in diversity. Others offer different explanations.

10. So, too, did the transparency and openness of the American goods and capital markets, stressed by Jeffry Frieden in "Capital Politics: Creditors and the International Political Economy," *Journal of Public Policy*, Vol. 8, Part 3/4, July–December 1988, pp. 265–86. Peter Cowhey notes that American political institutions and the electoral system in particular contributed significantly to the credibility of American promises about multilateralism. See Cowhey, "Elect Locally, Order Globally." (Paper prepared for Ford Foundation Conference on Multilateralism, San Diego, 1991.)

11. See Stephen Krasner, "United States Commercial and Monetary Policy: Unravelling the Paradox of External Strength and Internal Weakness," in Peter J. Katzenstein, ed., *Between Power and Plenty: Foreign Economic Policies of Advanced Industrial States* (Madison: University of Wisconsin, 1978), and Fred Block, *The Origins of International Economic Disorder: A Study of United States International Monetary Policy from World War II to the Present* (Berkeley: University of California Press, 1977).

12. See Cowhey, "Elect Locally."

13. For a recent discussion, see G. John Ikenber and C. Kupchan, "Socialization and Hegemonic Power," *International Organization*, Vol. 61, No. 8, Summer 1990, pp. 283–315.

14. Wayne Sandholtz proposed and initially developed these three "futures" as a means of focusing the argument. They draw also from Robert Gilpin, *U.S. Power and the Multinational Corporation: The Political Economy of Foreign Direct Investment* (New York: Basic Books, 1975).

15. For a theoretical discussion, see Duncan Snidal, "The Limits of Hegemonic Stability Theory," *International Organization*, Vol. 39, Autumn 1985.

16. See John Zysman and Stephen Cohen, "Double or Nothing: Open Trade and Competitive Industry," *Foreign Affairs*, Vol. 61, No. 5, Summer 1983, p. 1115.

17. *Ibid.*

18. Jagdish Bhagwati, *Protectionism* (Cambridge, Mass.: MIT Press, 1988).

19. But times change, and the truly important act of a great empire, as many would observe, is a gracious and peaceful descent from dominance.

20. For an earlier and insightful discussion, see Robert Gilpin, "An Alternative Strategy to Foreign Investment," in his *U.S. Power and the Multinational Corporation*.

21. *Ibid.*

22. Joan Pearce, "Subsidized Export Credit." (Paper presented at London Chatham House, Royal Institute of International Affairs, 1980.)

23. See Zysman and Cohen, "Double or Nothing," pp. 1115ff.

24. See also Michalski, *Interfutures* (Paris: OECD, 1979).

25. For a recent interpretation, see David Calleo and Claudia Morgenstern, eds., *Recasting Europe's Economies: National Strategies in the 1980s* (Lanham, Md.: University Press of America, 1990).

26. As Cohen and Zysman ("Double or Nothing") wrote almost a decade ago:

> It is becoming obvious that long-term conflict over national economic position and advantage underlies many of the present trade troubles. In the narrowest sense it is a question of which countries will create substantial commercial advantage in the growth industries of the future, which countries will be able to defend employment in today's mainline industries during that transition, and which countries will move up to substantial roles in traditional sectors. More broadly, the very international rules determining the appropriate roles for government in national and international economic life are being challenged and questioned. (p. 1113)

27. See John Zysman, "Trade, Technology, and National Competition," in Enrico Deiaco et al., eds., *Technology and Investment: Crucial Issues for the 1990s* (London: Pinter Publishers, 1990); and John Zysman, "Trade, Technology, and National Competition," *International Journal of Technology Management: Special Issue on Strengthening Corporate and National Competitiveness through Technology*, Vol. 7, Nos. 1/2, 1992.

28. HDTV is a good example. The issue here, as Chapter 1 argues, is not only about who assembles the televisions but also about who produces the semiconductors and the screens. By extension, the issue is also about the equipment makers who embody the production and product know-how. Those components and subsystems, and the production equipment to make them, are critical to diverse products that embed computer functions, from computers themselves through washing machines and jet fighters. Thus the debate about HDTV standards becomes an issue in the broader competition in advanced electronics technology.

29. See Laura Tyson and John Zysman, "The Politics of Productivity: Developmental Strategy and Production Innovation in Japan," in Chalmers Johnson, Laura Tyson, and John Zysman, eds., *Politics and Productivity: How Japan's Development Strategy Works* (New York: Ballinger, 1989).

30. Zysman, "Trade, Technology, and National Competition."

31. See John Zysman, *Governments, Markets, and Growth: Finance and the Politics of Industrial Change* (Ithaca, N.Y.: Cornell University Press, 1983), Chap. 1.

32. Traditional studies of economic growth usually find that a significant share of growth is attributable to technological development, but they treat the technology component of growth as exogenous or unexplained. For example, see Edward Denison, *Why Growth Rates Differ: Postwar Experience in Nine Western Countries* (Washington, D.C.: Brookings Institution, 1967). Technological change drives economic development, but formal economic theory treats technology as unchanging and, as a result, traditional models provide little help in understanding the longer-term evolution of technology or the economy. More precisely, technological change is exogenous to economic analysis, conceived as independent of the current allocation of resources.

33. Studies of the process and history of technical change indicate that technology is not a set of blueprints given by scientific advances that occur independently of the production process, but often is a joint output of the production process itself. The pace and direction of technological innovation and diffusion are shaped by production and market position.

34. A microchip does not, despite the belief of some, have the same economic value as a potato chip. We choose this example precisely because it was argued to one of us during the Reagan years by senior government officials that a dollar of grapefruit production was interchangeable with a dollar of semiconductor production.

35. Much of this argument was introduced to this debate by Stephen Cohen and was developed in Stephen Cohen and John Zysman, *Manufacturing Matters: The Myth of the Post-Industrial Economy* (New York: Basic Books, 1987).

36. *Ibid.* As an example, consider the relation between services and manufacturing. Some service activities, such as advertising, are downstream from production and will go on no matter who produces a good and where it is produced. These activities are not linked, or are very weakly linked to production. Other services, however, are more tightly bound to manufacturing, and some are very tightly linked indeed. If grain and animal production in Nebraska stopped and were replaced by imports, then veterinarians and other service workers would be displaced. In the traditional view, very tightly linked activities were bound by geography.

37. When the spillover is great and the knowledge from a sector is tacit and passes through community institutions, not markets, then that sector as a whole represents a piece of infrastructure to the economy. In some cases the same technology might pass through communities, remain tacit and non-market, or pass through markets more explicitly. If one country maintains an industry that is structured so that technological knowledge remains captive and is not readily available on the market, then other countries that would benefit from the technological spillovers from that industry must develop their own domestic producers. Technological externalities and the ability to capture them of course vary with industry and national organization. But the issue is as much one of political and social organization as of economic logic. The character of spillovers and the mechanisms states may use to confine them to a particular geography, such as restrictions on employment training or restraints in goods markets, are crucial to any arguments calling for technological autonomy or capacity as an element of development strategy.

38. Technology is malleable to some extent, and its particular form is set by the social molds in which it emerges. Particularly when a technology is in its infancy, the line of its evolution is fluid and inherently uncertain. An emphasis can be put on making steel stronger or lighter. As long as more than one direction of development is possible, not only the pace but also the direction of development is a matter of decision.

39. After that point, a new technology will not be developed unless it is so attractive that producers and users are willing (both economically and otherwise) to walk away from their investments in earlier technologies. If the gains from new technical approaches look marginal, they will be ignored; if gains look potentially important but slow to develop or very risky, they may never be captured.

40. See Harvey Sapolsky, *The Polaris System Development: Bureaucratic and Programmatic Success in Government* (Cambridge, Mass.: Harvard University Press, 1972). Some might see this as a form of redundancy. We would argue that it is not. A spare tire is redundant, but it is essential if there is a flat tire. A second phone line provides a cushion of capacity if the first one is in use. Both are identical to the apparatus they replace.

They are quite literally redundant, or extra, during ordinary conditions. Bets on a roulette wheel, however, are not identical; each is valuable precisely because it is different from the others. In terms of static efficiency, the extra or unused efforts would be duplications—wasted effort. In dynamic terms, the extra options are essential to guarantee success. Technology managers have often recognized this. Indeed, the Polaris submarine development program built multiple bets into the program at critical technological junctures. The biggest technical uncertainty was whether the missiles could be fired from below the surface, and a set of different projects was undertaken to solve the firing problem.

41. This argument is more fully elaborated in John Zysman, *Political Strategies for Industrial Order: State, Market and Industry in France* (Berkeley: University of California Press, 1977), and in Tyson and Zysman, "The Politics of Productivity."

One classic line of argument proposes that industrial organization (whether an industry is atomized, monopolized, or oligopolized—sometimes cartelized—or is somewhere in between) sets the framework within which the individual firm makes choices. Vary industry organization and firm behavior will change to include strategies for the use and development of technology. Although there is obvious merit to this argument, industry organization alone does not establish distinct technological trajectories, and organization itself may be influenced by other things. One of those deeper influences is the industrial/institutional structure of the economy—the structure of government and the character of institutions that regulate and arrange financial and labor markets as well as the markets for goods and services—that creates specific patterns of constraint and opportunity. Those patterns produce regularities in the strategies of governments and firms, which emerge quite visibly as industrial organization.

Let us contrast the French and Japanese cases. The pattern of French government interventionism has produced constant responses over the years. In the French model the government mobilized important resources—financial, institutional, and human—and directed them to specific tasks that were established in advance. The bases of this strategy are evident. The French bureaucratic system is centralized and could call on a financial system that was under the influence of the state. The industrial structure consisted of small firms that historically were protected from foreign threats and with competition often dampened at home. The French solution worked when the tasks at hand required mobilization of resources, when it was possible to define a limited number of technological results, and when the competitive market could be suppressed, controlled, or oriented by the state. The success was evident with Ariane, Airbus, the TGV, and the Minitel system. The strategy didn't work when the task was for a company to rapidly adapt its products and processes to changing international market conditions. As a result, the French position in consumer durables has always been weak, its position in electronic components untenable. In sum, a particular pattern of government policy in a specific industrial setting produced a particular pattern of technology development and trade.

Japanese interventionism produced a different pattern of market response. In Japan, the government acted as a gatekeeper during the postwar development years to break apart the multinational package of money, technology, and management. Government policy produced intense internal competition, structured so that pursuit of market share was the best way to pursue profits. Japanese firms responded by becoming "industrial followers" in expanding markets and by aggressively borrowing technology from abroad. Market share and expanded production led to rapid introduction of new generations of technology. Internal market logic which continued to be defined by state policy induced corporate responses that resulted in a production revolution, in a continuing corporate search for external technology and rapid product introduction, and in the interconnected web of production and distribution that so confounds foreigners. It also produced excess capacity in Japan that later spilled over into global markets in the form of dumping controversies. As in France, industrial organization was the outcome of a several-tiered process leading from politics to policy to market logic to corporate strategy and production innovation. The result in Japan was success in consumer durable sectors such as automobiles and consumer electronics; industrial organization contributed to a technology trajectory quite distinct from the French. The same institutional/industrial structure approach has been used to examine U.S.–Japanese competition in a series of industries. Again, the distinct strategies and technological trajectories produced a competitive interaction that favored the Japanese in many key sectors.

   Another approach would emphasize the character of social conflict and community or regional social organization in industry. Variation here, as in different industrial automation patterns among Fiat, Volvo, and Volkswagen, is explained by differences in the character of the conflict between labor and management in each of these corporations' countries. Fiat's approaches to automation, which truly displace labor, are certainly a response to the sabotage strikes and labor militancy of a generation ago, whereas automation in Japanese factories follows a very different logic, permitting the use of skilled labor as an experimental device. Another instance would be the two paths of development with the German machine tool industry, each with a distinct industrial organization, industry institutions, market strategy, and technological implications.

   A fourth line of argument distinguishes between the trajectories of defense and civilian high-technology industries in the United States. Here the defense trajectory is rooted in the nature of procurement, with essentially a single buyer purchasing from a limited set of final contractors and with products specified by performance, price and production cost are simply irrelevant. That market logic becomes entrenched in corporate strategy and organization with the result that military firms often do not have the organization skills or strategy to succeed in civilian markets. Indeed, a comparison of American and Japanese machine tool development traces a military and civilian line of technological evolution, each pushed by government policy.

   All of these arguments point to the same conclusion: The foundation of industrial structure includes features of the state and society that are not the same across the modern industrial economies. To take it one step further, multinational corporations and global financial institutions do not, in this view, sweep away the national foundations of trade, finance, and technology, and they do not create a homogenized single global market. The fact of MNCs and direct foreign investment does not mean that local variation disappears; instead, the MNC acts as an arbitrageur between these local markets. Multinationals certainly link the several national economies; they do not necessarily break them down.

   42. Learning curves, technological linkages, and nationally rooted technological developments all create the basis for firms in one nation to surge to advantage in world markets together, riding a common paradigm of technology emerging from their local experience.

   43. See Paul Krugman, "Increasing Returns, Monopolistic Competition, and International Trade," *Journal of International Economics*, Vol. 9, November 1979, pp. 469–79; "The New Theories of International Trade and the Multinational Corporation," in D. Audretsch and C. Kindleberger, eds., *The Multinational Corporation in the 1980s* (Cambridge, Mass.: MIT Press, 1982); Paul Krugman, *Rethinking International*

Trade (Cambridge, Mass.: MIT Press, 1990); and Elhanan Helpman and Paul Krugman, *Market Structure and Foreign Trade: Increasing Returns, Imperfect Competition, and the International Economy* (Cambridge, Mass.: MIT Press, 1985).

44. This section is drawn from work done jointly with Laura Tyson, Stephen Cohen, and Giovanni Dosi.

45. See Kozo Yamamura, *Economic Policy in Postwar Japan: Growth versus Economic Democracy* (Berkeley: University of California Press, 1967); Kozo Yamamura and Yasukichi Yasuba, eds., *The Political Economy of Japan* (Stanford, Calif.: Stanford University Press, 1987). See also Alice Amsden, *Asia's Next Giant: South Korea and Late Industrialization* (New York: Oxford University Press, 1989).

46. AnnaLee Saxenian, "In Search of Power: The Organization of Business Interests in Silicon Valley and Route 128," *Economy and Society*, Vol. 18, no. 1, February 1989. Gary Herrigel and Richard Locke, "Eppure si muove: The Political Economy of Industrial Change in Italy" (Ph.D. diss., MIT, 1989).

47. Indeed, although these arguments have limited circulation in the United States, they have a wider audience in Europe. Moreover, economic issues perceived as market problems in the United States are widely perceived as political issues in Europe. Also, in Japan technological development is more closely viewed as tied to political power. Contemporary trade disputes increasingly are argued on these terms among all of the pairs of advanced regions, not only those between the United States and Japan but also those between Europe and the United States and Europe and Japan. In Seattle, the home of Boeing, European aircraft policy is the focus. In Europe there is certainly concern with, and fascinated admiration for, Japanese development policy; but there is also concern with the commercial and development consequences of U.S. Department of Defense policies that promote particular sectors and restrict or retard the export of critical component technology to Europe.

48. See Tyson and Zysman, "The Politics of Productivity."

49. Paul Krugman's excellent analysis is a development of very similar arguments for a somewhat different set of concerns. It came to our attention as this chapter was being completed and we benefited from it in the final expression of these arguments. Paul Krugman, "The Move to Free Trade Zones." (Preliminary draft of a paper prepared for the Policy Implications of Trade and Currency Zones Symposium, Jackson Hole, Wyoming, August 22–24, 1991.)

50. Peter Morici's work is the clearest on this subject. See *U.S. Economic Policies Affecting Industrial Trade: a Quantitative Assessment* (Washington, D.C.: NPA Committee on Changing International Realities, 1983); *The Global Competitive Struggle: Challenges to the U.S. and Canada* (Washington, D.C.: Canadien and American Committee, 1987); *Meeting the Competitive Challenge: Canada and the U.S. in the Global Economy* (Washington, D.C.: Canadien and American Committee, 1988).

51. See note 50 above.

52. See note 50 above.

53. For example, see Paul Krugman, "Introduction: New Thinking about Trade Policy," in Paul Krugman, ed., *Strategic Trade Policy and the New International Economies* (Cambridge, Mass.: MIT Press, 1986).

54. We thank Burgess Laird for conversations on this topic.

55. The following section draws from work that was originally presented in the journal *International Security* by Stephen Van Evera, Jack Snyder, Scott Sagan, and others. See Steven E. Miller, ed., *Military Strategy and the Origins of the First World War* (Princeton, N.J.: Princeton University Press, 1985).

56. See Jack Snyder, *The Ideology of the Offensive: Military Decision-Making and the Disasters of 1914* (Ithaca, N.Y.: Cornell University Press, 1984).

57. For example, George Gilder among others argues that emerging microprocessor technologies actually favor the American model of industrial organization over the Japanese, reversing the material advantages that our competitors have had over the past decade in the production of standardized commodity chips in large batches at low cost. Although we suspect that Gilder is wrong—that microprocessors will themselves take on commodity characteristics and that segments of the market will be dominated by those with volume production technology—he may be right in some objective sense, and it may not matter to this argument.

58. Jagdish Bhagwati, *Protectionism* (Cambridge, Mass.: MIT Press, 1988).

59. For a somewhat different route to a similar conclusion, see John Mueller, *Retreat from Doomsday: The Obsolescence of Major War* (New York: Basic Books, 1989). Also see Carl Kaysen, "Is War Obsolete?" *International Security*, Vol. 14, Spring 1990.

60. See Barry Buzan, "Economic Structure and International Security: The Limits of the Liberal Case," *International Organization*, Vol. 38, Autumn 1984.

61. There were scattered reports during the recent Gulf War of shortages for components made in Japan of certain United States weapons systems. Regardless of the accuracy of those reports or even their specific impact, the perception is likely to be important.

62. See *New York Times*, May 22, 1991, pp. C1, 7. Japanese television depends not on cables but on a more technically sophisticated network of satellites that beam programs directly to consumers with small reception devices.

63. For a full discussion, see Steve Weber, *Cooperation and Discord in U.S.–Soviet Arms Control* (Princeton, N.J.: Princeton University Press, 1991).

Epilogue

1. Italy's problem with Albanian immigrants is just the beginning of migration problems that will plague Europe for years to come.

2. The problems in Yugoslavia will find ample parallel in Kazakhstan and many other of the Soviet republics.

3. The reference is, of course, to Francis Fukuyama's thoughtful, oft-misconstrued paean to the end of the Cold War, in Francis Fukuyama, "The End of History," *The National Interest*, No. 16, Summer 1989, pp. 3–18.

4. *Forward pricing* occurs when a firm prices below current costs in anticipation of generating sufficient demand to push actual production costs down below the price target. *Dumping* is selling below a fair market value.

5. For a thorough and thoughtful explication of the case for and against managed trade, see Laura D'Andrea Tyson, *Who's Bashing Whom: Trade Conflicts in High Technology Industries* (Washington, D.C.: Institute of International Economics, forthcoming 1992).

6. See the various works of BRIE on this subject as well as the works of the MIT Commission on Industrial Productivity.

Index

251

The
Highest
Stakes